T0324514

Oscillator and Pendulum with a Random Mass

Oscillator and Pendulum with a Random Mass

Moshe Gitterman
Bar-Ilan University, Israel

 **World Scientific**

NEW JERSEY · LONDON · SINGAPORE · BEIJING · SHANGHAI · HONG KONG · TAIPEI · CHENNAI

Published by

World Scientific Publishing Co. Pte. Ltd.

5 Toh Tuck Link, Singapore 596224

USA office: 27 Warren Street, Suite 401-402, Hackensack, NJ 07601

UK office: 57 Shelton Street, Covent Garden, London WC2H 9HE

British Library Cataloguing-in-Publication Data

A catalogue record for this book is available from the British Library.

ISBN 978-981-4630-74-0

In-house Editor: Song Yu

Typeset by Stallion Press
Email: enquiries@stallionpress.com

Printed in Singapore

Preface

The first step in solving scientific or engineering problem is the choice of a simplified model. The simplest, the most general and the most widely used is that of the harmonic oscillator, where the force acting on a particle is directly proportional to its displacement. This model is used for the description of many different phenomena in physics, chemistry, biology and social sciences [1]. However, this model requires improvement in at least two directions. Most processes are nonlinear, and the simplest model for their description is a pendulum. The pendulum is modelled as a massless rod with the point mass on its end. The second restriction of an oscillator model consists in its pure mechanical nature (zero temperature), and in order to describe real phenomena, one has to add random forces to the oscillator equation. These sources of noises may have internal origin or external origin, and they may enter the oscillator equation additively or multiplicatively. Additive noise arises from non-zero temperatures or from very rapid dynamics of certain degrees of freedom. Multiplicative noise arises from the stochastic nature of external fields and boundary conditions.

The oscillator or pendulum equations may contain random frequency, random damping or a random mass. The first two cases have been intensively studied, and we describe them in the Introduction. The main aim of this book is the detailed analysis of the oscillator and pendulum with a random mass. A simple physical example is the Brownian motion with adhesion. The usual Brownian motion describes a particle subject to dynamic systematic force, proportional to velocity, and a random force exerted on the Brownian particle by the molecules of the surrounding medium. The fluctuation force comes from the different number of molecular collisions with a Brownian particle from two opposite sides resulting in its random motion. This phenomenon is described by a differential equation with additive random force. Brownian motion with adhesion means that the molecules of the surrounding medium not only randomly collide with the Brownian particle, which produce its well-known zigzag motion, but they also stick to the Brownian particle for some (random) time, thereby changing its mass, which is described by an additional multiplicative random force in the dynamic equation.

The organization of this book is as follows. The introductory part contains the general description of oscillators and pendulums as well as of all four types of randomness in the oscillator and pendulum equation. For three of them (additive noise, random frequency and random damping) the first two moments are calculated. The main part of the book contains the detailed analysis of an oscillator with a random mass. The final part of the book contains the analysis of a pendulum with a random mass. We expect that the model of an oscillator and a pendulum with a random mass will find many applications in science and in technology.

Contents

Chapter 1

Introduction

A particle of mass m which is displaced from its equilibrium position at $x = 0$ is subject to a restoring attractive force and a viscous force which, in first approximation, are proportional to the displacement and velocity, respectively. Thus, for an one-dimensional system, Newton's law gives

$$m\frac{d^2x}{dt^2} + 2\gamma\frac{dx}{dt} + \omega^2 x = 0 \tag{1}$$

where $2\gamma/m$ is the damping rate and $\omega/m^{1/2}$ is the intrinsic frequency. The solution of this equation is

$$x = \left[\frac{\gamma x_0 + m y_0}{m\omega_1}\sin(\omega_1 t) + x_0\cos(\omega_1 t)\right]\exp\left(-\frac{\gamma t}{m}\right) \tag{2}$$

where

$$\omega_1 = \frac{\left(m\omega^2 - \gamma^2\right)^{1/2}}{m}; \quad x_0 = x(t = 0); \quad y_0 = \frac{dx}{dt}(t = 0) \tag{3}$$

If the harmonic oscillator is subject to an external periodic force $A\exp(i\Omega t)$,

$$m\frac{d^2x}{dt^2} + 2\gamma\frac{dx}{dt} + \omega^2 x = A\exp(i\Omega t) \tag{4}$$

then the complete solution of this equation consists of solution (2) of the homogeneous equation (1) and the following solution of the non-homogeneous equation,

$$x = \frac{A\exp(i\Omega t)}{\omega^2 - m\Omega^2 + 2i\gamma\Omega} \tag{5}$$

The last equation indicates the possibility of a (dynamic) resonance, when the external frequency Ω approaches the intrinsic frequency $\omega/m^{1/2}$.

1.1 Harmonic oscillator with external noise

In the foregoing we considered a pure mechanical system (zero temperature). For non-zero temperatures, the dynamic equation (1) has to be supplemented by the thermal noise

$$m\frac{d^2x}{dt^2} + 2\gamma\frac{dx}{dt} + \omega^2 x = \eta(t) \qquad (6)$$

where $\eta(t)$ is the random variable with zero mean, $\langle\eta(t)\rangle = 0$, and variance $\langle\eta^2(t)\rangle$, which for thermal noise satisfies the fluctuation-dissipation theorem $\langle\eta^2(t)\rangle = 4\gamma\kappa T$, where κ is the Boltzmann constant [2]. The latter simply means that the power entering the system from the external force must be entirely dissipated and given off to the thermostat in order that the equilibrium state of the system not be disturbed. Another way to justify the validity of Eq. (6) is as follows: in considering only one (slow) mode $x(t)$ of a complex system, one may take into account the influence of other (fast) modes by introducing a random force into dynamic equation with no special requirements for the value of $\langle\eta^2(t)\rangle$.

Upon averaging Eq. (6), one finds that the first moment is given by Eq. (2) while the second moment $\langle x^2(t)\rangle$ for white noise $\langle\eta^2(t)\rangle = D$ is defined by the variance,

$$[x(t) - \langle x(t)\rangle]^2$$
$$= \frac{D}{2\gamma\omega 2}\left\{1 - \left[1 + \frac{\gamma}{\omega_1}\left(\sin 2\omega_1 t + \frac{2\gamma}{\omega_1}\sin^2\omega_1 t\right)\exp(-2\gamma t)\right]\right\} \qquad (7)$$

For $m = 1$, dichotomous noise of strength σ^2 and inverse correlation length λ, the variance reaches the following stationary $(t \to \infty)$ value

$$\lim_{t\to\infty}[x(t) - \langle x(t)\rangle]^2 = \frac{\sigma^2(\lambda + 2\gamma)}{2\omega^2\gamma(\lambda^2 + 2\lambda\gamma + \omega^2)} \qquad (8)$$

which for white noise is reduced to

$$\lim_{t \to \infty} [x(t) - \langle x(t) \rangle]^2 = \frac{D}{2\omega^2 \gamma} \qquad (9)$$

Both equations (7) and (8) were obtained already in 1945 [3].

1.2 Ito-Stratonovich dilemma

In spite of the fact that the stochastic differential equations were introduced more than a hundred years ago, there are still many interesting problems concerning these equations. We consider here the Ito-Stratonovich dilemma, using the generalized Schlogl model as an example [4], which is described by the following equation,

$$\frac{dx}{dt} = -x^3 + \beta x + |x|^\nu \xi(t) \qquad (10)$$

where ν is an arbitrary number and $\xi(t)$ is white noise with the correlator

$$\langle \xi(t_1) \xi(t_2) \rangle = 2D\delta(t_1 - t_2) \qquad (11)$$

The Langevin equation (10) is not completely defined due to the Ito–Stratonovich dilemma, namely, it is not clear which value of t one has to insert in the δ-function (11) and afterwards in the probability distribution. The two possibilities are: before the jump (Ito) or the averaged of before and after the jump (Stratonovich). This choice is very important since it leads to different Fokker-Planck equation for the probability distribution $P(x,t)$ [5]. The first is described by the dynamic equation of the form

$$\frac{\partial P}{\partial t} = -\frac{\partial}{\partial x}[(-x^3 + \beta x)P] + D\frac{\partial^2}{\partial x^2}[|x|^{2\nu}P] \qquad (12)$$

and the second by the equation

$$\frac{\partial P}{\partial t} = -\frac{\partial}{\partial x}[(-x^3 + \beta x)P] + D\frac{\partial}{\partial x}\left\{|x|^{2\nu}\frac{\partial}{\partial x}[|x|P]\right\} \qquad (13)$$

Another important factor which defines the behavior of a system is the value of ν.

1. For $\nu = 0$, there are three steady states for $\beta > 0$, namely, $x_0 = 0$ (unstable) and $x_0 = \pm\sqrt{\beta}$ (stable), and one stable state for $\beta < 0$, namely, $x_0 = 0$.
2. For $\nu > 0$, multiplicative noise has an attractive effect, i.e., it attracts the probability $P(r, t)$ to unstable steady state, whereas for $\nu < 0$, the noise term makes system more stable.
3. The stationary probability distribution $P(x)$ (at $t \to \infty$) is defined by the competition between the noise and the damping terms in Eq. (10), which define ingoing and outgoing energy, respectively, i.e., by three numbers, ν, β and D. In the Stratonovich approach [4] $P(x) = C \exp\left[-V(x)/D\right]$, where C is the normalization constant and

$$P(x) = \frac{x^{4-2\nu}}{4 - 2\nu} - \frac{\beta x^{2-2\nu}}{2(1 - \nu)} + \nu D \lg x; \quad \text{for } \nu \neq 1 \quad \text{and} \quad \nu \neq 2$$

$$P(x) = \frac{x^2}{2} + (D - \beta) \qquad\qquad \text{for } \nu = 1$$

$$P(x) = (1 + 2D) \ln x + \frac{\beta}{2x^2} \qquad \text{for } \nu = 2$$

$$\tag{14}$$

For the Ito approach, one has to replace νD by $2\nu D$ in (14). When $\nu = 1$, the probability vanishes as $x \to \infty$ for $\beta > D$ and $\beta < D$. The main difference between these two approaches occurs for $D < \beta < 2D$, where for the Ito case, the probability diverges as $x \to 0$, whereas for the Stratonovich case the probability becomes zero.

Since the Ito and the Stratonovich approaches sometimes lead to different results, one has to be careful which to apply to the analysis of real systems.

1.3 Harmonic oscillator with random frequency

The random force $\eta(t)$ enters Eq. (6) additively. When the noise has an external origin rather than an internal origin, for instance if it arises from the fluctuation of the potential energy

$U = \omega^2 x^2[1 + \xi(t)]/2$, the equation of motion of an oscillator with an external periodic force will take the following form,

$$m\frac{d^2x}{dt^2} + 2\gamma\frac{dx}{dt} + [1 + \xi(t)]\omega^2 x = A\exp(i\Omega t) \qquad (15)$$

One additional remark is necessary when dealing with multiplicative noise. It is not clear at which time t the value of $x(t)$ has to be taken in the $\xi(t)x(t)$ term in Eq. (15) with respect to the characteristic times of $\xi(t)$ (the so-called Ito-Stratonovich dilemma). In the following we use the Stratonovich interpretation.

The internal dynamics of a force-free ($A = 0$) harmonic oscillator with random frequency is a subject that have been extensively investigated in different fields including physics (off-on intermittence [6], dye lasers [7], wave propagation in random media [8]), biology (population dynamics) [9], economics (stock market prices) [10], and so on.

In order to find the first two moments, one has to rewrite Eq. (15) as two first-order differential equations, multiply them by $\xi(t)$ and average, leading to four equations for $\langle x\rangle, \langle y\rangle, \langle\xi x\rangle$ and $\langle\xi y\rangle$ [11]. Equation (15) without external force, $A = 0$, has been extensively studied [12]. It turns out that the white noise fluctuations in frequency do not affect the first moment, whereas for dichotomous noise, the equation for $\langle x\rangle$ has the following form,

$$\frac{d^2\langle x\rangle}{dt^2} + (2\gamma - \omega^2 c_1)\frac{d\langle x\rangle}{dt} + \omega^2(1 - \omega c_2)\langle x\rangle = 0 \qquad (16)$$

where

$$c_1 = 2\int_0^\infty \langle\xi(t)\xi(t-\tau)\rangle[1 - \cos(2\omega\tau)]d\tau;$$

$$c_2 = 2\int_0^\infty \langle\xi(t)\xi(t-\tau)\rangle\sin(2\omega\tau)d\tau \qquad (17)$$

In the presence of an external force $A\exp(i\Omega t)$ in Eq. (15), its averaged solution is $\langle x\rangle = a\sin(\Omega t + \phi)$, where

$$a = \left[\frac{f_1^2 + f_2^2}{f_3^2 + f_4^2}\right]^{1/2} ; \quad \tan\phi = \left(\frac{f_1 f_3 + f_2 f_4}{f_1 f_4 - f_2 f_3}\right) \qquad (18)$$

with

$$f_1 = 2A\Omega(\lambda + \gamma); \quad f_2 = A(\Omega^2 - \omega^2 - \lambda^2 - 2\lambda\gamma);$$

$$f_3 = (\Omega^2 - \omega^2)(\Omega^2 - \omega^2 - \lambda^2) - \sigma^2 - \Omega^2(6\lambda\gamma + 4\gamma^2) + 2\lambda\omega^2;$$

$$f_4 = \Omega(\lambda + 2\gamma)[2(\omega^2 - \Omega^2) + 2\lambda\gamma] \tag{19}$$

The second moment $\langle x^2 \rangle$ for the oscillator with random frequency can be found in the same way as described below for random damping [11], and we do not bring these results here.

1.4 Harmonic oscillator with random damping

Another possibility for the generalization of the dynamic equation (4) is the incorporation of random damping,

$$m\frac{d^2x}{dt^2} + 2\gamma[1 + \xi(t)]\frac{dx}{dt} + \omega^2 x = A\sin(\Omega t) + \eta(t) \tag{20}$$

The first time this equation (with $A = 0$) was used for the problem of water waves influenced by a turbulent wind field [13]. However, this equation (with the coordinate x and time t replaced by the order parameter and coordinate, respectively) transforms into the Ginzburg-Landau equation with a convective term, which describes phase transitions in moving systems [14], which is, in fact, a new field of research. There are an increasing number of problems, which can be investigated by the methods under study, where the particles are advected by the mean flow. These include problems of phase transition under shear [15], open flows of liquids [16], Rayleigh-Benard and Taylor-Couette problems in fluid dynamics [17], dendritic growth [18], chemical waves [19], and the motion of vortices [20].

In order to find the first moment $\langle x \rangle$, one has to rewrite Eq. (20) as two first-order differential equations,

$$\frac{dx}{dt} = y; \quad \frac{dy}{dt} = -2\gamma y - 2\gamma\xi y - \omega^2 x + A\sin(\Omega t) + \eta(t) \tag{21}$$

After averaging, equations (21) contain a new correlator $\langle \xi y \rangle$, which has to be found separately. Multiplying the second equation in (21) by ξ and averaging, one gets two new correlators, $\langle \xi x \rangle$ and $\langle \xi^2 y \rangle$. The latter can be found by the splitting procedure, $\langle \xi(t_1)\xi(t_2)y(t_2)\rangle = \langle \xi(t_1)\xi(t_2)\rangle\langle y(t_2)\rangle$. Finally one obtains a system of four equations for the four variables, $\langle x\rangle, \langle y\rangle, \langle \xi x\rangle$ and $\langle \xi y\rangle$, the solution of which is $\langle x(t)\rangle = a\sin(\Omega t + \phi)$, where

$$a^2 = \frac{A^2(f_2^2 + \Omega^2\lambda_2^2)}{(2\Omega^2\gamma\lambda_2 - f_1f_2 - 4\gamma^2\Omega^2\sigma^2)^2 + 4\Omega^2[\lambda_1(\gamma\lambda - f_1) - 2\lambda\gamma^2\sigma^2]^2}$$

(22)

and

$$\tan\phi = \frac{2\Omega\gamma f_2^2 + 4\Omega^3\gamma^2\lambda_2 - 4\Omega\gamma^2\sigma^2(\lambda_1\Omega^2 + \lambda\omega^2 + \lambda^2\lambda_1)}{f_1(f_2^2 + \Omega^2\lambda_2^2) + 4\Omega^2\gamma^2\sigma^2(f_1 + \lambda^2)} \quad (23)$$

where

$$\lambda_1 = \lambda + 2\gamma; \quad \lambda_2 = 2(\lambda + \gamma); \quad f_1 = \Omega^2 - \omega^2; \quad f_2 = f_1 - \lambda\lambda_1$$

(24)

As initial conditions, the non-zero stationary $(t \to \infty)$ second moment $\langle x^2\rangle$ requires an additional additive noise for the field-free case $A = 0$ while for the driven oscillator there is no need for the additive noise, and we put $\eta(t) = 0$. The additive white noise is uncorrelated with the multiplicative noise.

One multiplies Eqs. (21) by $2x$ and $2y$ and averages, and then multiplies by y and x and averages the sum of these equations. In the three resulting equations, one gets for white noise $\eta(t)$, $\langle y\eta(t)\rangle = D$ and $\langle x\eta(t)\rangle = 0$. In addition, these equations contain new correlators $\langle \xi y^2\rangle$ and $\langle \xi xy\rangle$. One can calculate these and the analogous correlator $\langle \xi x^2\rangle$ using the Shapiro-Loginov procedure, getting [11] three additional equations for six variables, $\langle x^2\rangle$, $\langle y^2\rangle$, $\langle xy\rangle$, $\langle \xi x^2\rangle$, $\langle \xi y^2\rangle$ and $\langle \xi xy\rangle$.

$$\frac{d}{dt}\langle x^2\rangle = 2\langle xy\rangle;$$

$$\frac{d}{dt}\langle y^2\rangle = -4\gamma\langle y^2\rangle + 16D\gamma^2\langle y^2\rangle - 2\omega^2\langle xy\rangle + 2A\langle y\rangle\sin(\Omega t);$$

$$\frac{d}{dt}\langle xy\rangle = \langle y^2\rangle - 2\gamma\langle xy\rangle + 4D\gamma^2\langle xy\rangle - \omega^2\langle x^2\rangle + A\langle x\rangle\sin(\Omega t)$$

(25)

1.5 Harmonic oscillator with multiplicative and additive noise

So far, we have considered an oscillator with one random force. However, many interesting phenomena occur when there are two random forces. As an example, we consider the overdamped (large damping) first-order differential equations with two sources of multiplicative noise [21], linear and quadratic multiplicative noise [22], linear multiplicative and quadratic additive noise [23].

1.5.1 *Two stochastic resonances in a system with two sources of multiplicative noise*

The phenomenon of stochastic resonance, which we will consider in detail for an oscillator with a random mass in Section 2.20, has aroused considerable interest due to its many applications in science and technology. This counter-intuitive phenomenon of the amplification of a small external signal by noise, which contradicts the intuition, has already found many applications [24].

One of the models is an electric circuit with electric constant inductance L, constant voltage U_0 and an oscillatory voltage $A\sin(\omega t)$ with two dichotomous resistors $\xi_1(t)$ and $\xi_2(t)$. The stochastic dynamic equation of motion for the current density I has the following form,

$$L\frac{dI}{dt} + [\xi_1(t) + \xi_2(t)]\,I = U_0 + A\sin(\omega t) \qquad (26)$$

The asymmetric dichotomous noise $\xi_1(t)$ fluctuates between two values a and b with mean waiting time t_a and t_b in each of these two states. For the second noise $\xi_2(t)$, the associated parameters are c, d, t_c and t_d. The stationary first moments of the noise and their correlation functions are [5]

$$\langle \xi_1(t) \rangle = \frac{ak_1 + b}{1 + k_1} \equiv g_1; \quad \langle \xi_2(t) \rangle = \frac{ck_2 + d}{1 + k_2} \equiv g_2;$$

$$\langle \xi_{1,2}(t_1)\xi_{1,2}(t_2) \rangle = D_{1,2}\lambda_{1,2}\exp[-\lambda_{1,2}(t_1 - t_2)] + \langle \xi_{1,2}(t) \rangle^2;$$

$$\langle \xi_1(t)\xi_2(t) \rangle = g_1 g_2 \qquad (27)$$

where

$$k_1 = t_a/t_b, \quad k_2 = t_c/t_d, \quad \lambda_1 = (1 + k_1)\tau_b, \quad \lambda_2 = (1 + k_2)\tau_d;$$

$$D_1 = \frac{k_1(a - b)^2}{\lambda_1(1 + k_1)^2}; \quad D_2 = \frac{k_2(c - d)^2}{\lambda_2(1 + k_2)^2} \tag{28}$$

After cumbersome calculations, one obtains for the first moment of the current, $\langle I(t) \rangle$ [21]

$$\langle I(t) \rangle = A_1 + M \sin(\omega t + \phi) \tag{29}$$

where A_1 and M are known functions of the noise parameters. The graph of the eigenvalues of Eq. (26) as a function of the strength of the noise, which displays the stochastic resonance phenomenon, shows that there are two stochastic resonances for the amplitude of the periodic response to two sources of noise, $\xi_1(t)$ and $\xi_2(t)$. This result is supported by numerical calculations [21]. In addition to multiplicative noise, considered here, stochastic resonances also appear for two sources of noise, one additive and one multiplicative ("double stochastic resonance") [25].

1.5.2 *Multiplicative linear and quadratic noise*

The specific problem considered in this section is the combination of linear and quadratic multiplicative color noise acting on a linear overdamped harmonic oscillator forced by a periodic term,

$$\frac{dx}{dt} = -[a_0 + a_1\xi(t) + a_2\xi^2(t)]x + A\sin(\Omega t) \tag{30}$$

where

$$\langle \xi_1(t)\xi_2(t) \rangle = \alpha D \exp(-\alpha|t_1 - t_2|) \tag{31}$$

with intensity of noise D and correlation time α^{-1}.

Let us start with the simple case $A = a_2 = 0$. Then, the averaged solution of Eq. (30) has the following form,

$$\langle x \rangle = \langle x_0 \rangle \exp[-(a_0 - Da_1^2)t] \exp\left\{-\frac{Da_1^2}{\alpha}[1 - \exp(-\alpha t)]\right\} \tag{32}$$

In the long-time limit, $\alpha t \gg 1$, the time dependence of $\langle x \rangle$ does not depend on the color of the noise α. It depends only on the intensity

D of the noise so that for $D < D_c = a_0/a_1^2$, $\lim_{t\to\infty}\langle x\rangle = 0$ and for $D > D_c$, $\langle x\rangle$ diverges. This is the simplest example of noise-induced instabilities. If in the initial system (30) $A = a_1 = 0$, then for $a_0 > 0$ (stable system), the noise does not influence the stability of the system while for $a_0 < 0$ (unstable system), the noise can stabilize the system (noise induced stabilization).

The more general case of Eq. (30) with $A = 0$ is considered in detail in [26]. This publication describes some interesting properties of a harmonic oscillator with squared multiplicative noise. For low noise intensity ($D < D_1$), the system is stable. The increase in D leads to the destabilization of the system. However, the system becomes stable again (!) for larger noise intensity. For the solution of the full equation (30), the authors of [22] solved for $x(t)$, and for averaging $\exp\{-\int_{t_1}^{t}[a_0+a_1\xi(\tau)+a_2\xi^2(\tau)]d\nu\}$, they used a functional integral representation. The final result for the stationary state averaged first moment is

$$\langle x(t)\rangle = C\sin(\Omega t - \phi) \tag{33}$$

where

$$C = \frac{A}{\sqrt{2ca_2/\alpha}}\sqrt{B_1^2 + B_2^2}; \quad \phi = \arctan\left(\frac{B_1}{B_2}\right) \tag{34}$$

with B_1 and B_2 being some combination of the original parameters.

1.5.3 *Stochastic resonance in an overdamped harmonic oscillator driven by multiplicative dichotomous and additive quadratic noise*

We bring here the results of the analysis of an overdamped oscillator with random frequency and additive quadratic noise, which is described by the following equation

$$\frac{dx}{dt} = -[c + \xi(t)]x + A\cos(\Omega t) + \eta^2(t) \tag{35}$$

where $\xi(t)$ and $\eta(t)$ are sources of non-correlated dichotomous noise,

$$\langle \xi(t_1)\xi(t_2)\rangle = \sigma_1\exp[-\lambda|t_1 - t_2|];$$
$$\langle \eta(t_1)\eta(t_2)\rangle = \sigma_2\exp[-\lambda|t_1 - t_2|] \tag{36}$$

As in Section 1.5.1, we assume that the asymmetric dichotomous noise $\xi_1(t)$ fluctuates between two values a and b with the mean waiting time t_a and t_b in each of these two states. For the second noise $\eta(t)$, the associated parameters are c, d, t_c and t_d. Using the Shapiro-Loginov procedure, the authors of [23] found the asymptotic values $(t \to \infty)$ of the first moment (x) and the second moment $\langle x^2 \rangle$,

$$\langle x \rangle = A \frac{f_1 \cos(\Omega t) + f_2 \sin(\Omega t)}{f_3} + f_4 \tag{37}$$

$$\langle x^2 \rangle = \frac{1}{c(2c + 2\Lambda_1 + \lambda) - 2\sigma_1}$$

$$\times \left\{ \frac{\sigma_2 \Lambda_2^2 [2(c + \Lambda_1 + \lambda)^2 + (c + \Lambda_1)(+2\sigma_1)]}{(c + \lambda)(c + \Lambda_1 + 2\lambda)} \right.$$

$$+ \frac{A^2 [f_1(2c + 2\Lambda_1 + \lambda) - 2\sigma_1(\Omega^2 - d_1 d_2)]}{2f_3}$$

$$\left. + \sigma_2 \left[f_4(2c + 2\Lambda_1 + \lambda) + \frac{2\sigma_1 \sigma_2}{d_1 d_2} \right] \right\} \tag{38}$$

where

$$f_1 = c\Omega^2 + (c + \Lambda_1 + \lambda)d_1 d_2; \quad f_2 = c[\Omega^2 + (c + \Lambda_1 + \lambda)^2 + \sigma_1]$$

$$f_3 = (\Omega^2 + d_1^2)(\Omega^2 + d_2^2);$$

$$f_4 = \frac{\sigma_2}{d_1 d_2}(c + \Lambda_1 + \lambda)d_{1,2} = c + \frac{\lambda + \Lambda_1}{2}$$

$$\pm \sqrt{\frac{(\lambda + \Delta_1)^2}{4} + \sigma_1}$$

$$\Lambda_1 = a - b; \quad \Lambda_2 = c - d: \quad \lambda = \tau_c + \nu_d; \quad \sigma_1 = ab; \quad \sigma_2 = cd \tag{39}$$

Equation (37) shows the dependence of the output-input ratio as a function of the oscillator and random frequencies, the noise correlation time and the multiplicative noise intensity. All these results show resonant type behavior [23], and the non-monotonic dependence on the output-input ratio of the multiplicative noise σ_1, which with increasing σ_1, first decreases, then increases, and finally decreases again.

1.6 Harmonic oscillator with a random mass

We recently studied [27] another possibility for introducing randomness in the oscillator equation (6) by considering an oscillator with a random mass. The appropriate equation of motion has the following form

$$m[1 + \xi(t)]\frac{d^2x}{dt^2} + 2\gamma\frac{dx}{dt} + \omega^2 x = \eta(t) \qquad (40)$$

This book is devoted to the analysis of an oscillator with random mass. Here we only note that there are many situations in which chemical and biological systems contain small particles, which are capable not only of colliding with a large particle, but they may also adhere to it, thereby changing its mass. The essential difference between this case and the two previous cases lies in the fact that for random frequency and random damping, one can use the simplest form of random noise $\xi(t)$ — white noise with the correlation function $\langle\xi(t_1)\xi(t_2)\rangle = D\delta(|t_1 - t_2|)$ whereas for a random mass, due to the positivity of mass, the simplest form of $\xi(t)$ is dichotomous noise,

$$\langle\xi(t_1)\xi(t_2)\rangle = \sigma^2 \exp[-\lambda|t_1 - t_2|] \qquad (41)$$

1.7 Pendulum

The harmonic oscillator is the simplest linear model. However, almost all phenomena in Nature are nonlinear. The simplest nonlinear generalization of the oscillator model is a pendulum. Recently we considered [28] a stochastic oscillator with random mass, where the surrounding molecules are able not only to collide with the oscillator, but also adhere to it, randomly changing its mass. Here we consider the same idea for a pendulum.

The simplest model of a pendulum is a material point (bob) of mass m suspended by a massless rod of length l and able to rotate around the rigidly attached the end of the rod. The position of the bob is defined by the angle ϕ of the rod with respect to its vertical position. The driving moment of the bob, induced by gravitation, $mgl\sin\phi$, is balanced by the rotary moment $ml^2 d\phi/dt$. Therefore, the mass of the bob m is eliminated from the equation of motion for the angle ϕ, and the frequency or period of bob rotation does not depend on the bob mass. The constants A and α of the solutions $\phi = A \sin(t + \alpha)$ of the second-order differential equation

$$ml^2 \frac{d^2\phi}{dt^2} = -mgl \sin\phi \tag{42}$$

are determined from the initial conditions $\phi(t = 0) = d\phi/dt$ ($t = 0) = 0$.

The energy of the bob is

$$E = \frac{ml^2}{2} \left(\frac{d\phi}{dt}\right)^2 - mgl \cos\phi \tag{43}$$

One can express the constants of integration in terms of the maximum angle β of the deviation from the vertical position of the rod, where the energy E is equal to $mgl \cos\beta$. Then

$$\frac{ml^2}{2} \frac{d\phi}{dt} = mgl(\cos\phi - \cos\beta) \tag{44}$$

and the constants of integrations are E and β. If the pendulum motion is damped, where the damping is proportional to the angular velocity $d\phi/dt$, the equation of motion (42) has an additional term $\gamma d\phi/dt$,

$$\frac{d^2\phi}{dt^2} + \frac{\gamma}{ml^2} \frac{d\phi}{dt} + \frac{g}{l} \sin\phi = 0 \tag{45}$$

In this case, the bob mass m does not drop out of the equation. For $t \to \infty$, the pendulum hangs in the downward position, $\phi = 0$.

In the presence of random collisions with the surrounded molecules, Eq. (45) takes the following form

$$\frac{d^2\phi}{dt^2} + \frac{\gamma}{ml^2} \frac{d\phi}{dt} + \frac{g}{l} \sin\phi = \eta(t) \tag{46}$$

This equation describes the energy balance: the power entering the system from the external random force must be entirely dissipated and transferred to the thermal bath in order that the equilibrium state of the system should not be disturbed. Another way to justify Eq. (46) is as follows. When considering only one slow mode $u(t)$ of a complex system, one may take into account the influence of other (fast) modes by introducing into the dynamic equation for the slow mode a random force, which we consider as white noise with correlator $\langle \eta(t_1)\eta(t_2)\rangle = D\delta(T_2 - t_1)$. The exact balance (steady state of a pendulum) is determined by the fluctuation-dissipation theorem, $D = \kappa T\gamma$. The stationary Fokker-Planck equation for the probability function $P(\phi, d\phi/dt = \Omega)$, which corresponds to the Langevin equation (46), is

$$\Omega \frac{\partial P}{\partial \phi} = \frac{\partial}{\partial \Omega}\left\{\left[\frac{\gamma}{2}\Omega + \frac{g}{2l}\sin\phi\right]P\right\} + \frac{D}{2}\frac{\partial^2 P}{\partial \Omega^2} \qquad (47)$$

It is easy to check that the solution of Eq. (47) has the following form

$$P(\phi, \Omega) = C\exp\left(-\frac{\gamma}{2D}\Omega^2\right)\exp\left(-\frac{\gamma g}{lD}\sin^2\phi\right) \qquad (48)$$

There are two types of pendulum motion, depending on the average angular velocity $d\phi/dt = \Omega$,

$$\langle\Omega\rangle = \int_0^\infty \Omega\exp\left[-\frac{\gamma}{2D}\Omega^2\right]d\Omega = \left(\frac{\pi D}{2\gamma}\right)^{1/2} \qquad (49)$$

which are defined by the ratio of the energies supplied by the noise and escaped the system through the damping. For small ratio D/γ, the pendulum will oscillate around its stable position, whereas for large D/γ, the motion will be circular. The average square angle ϕ is given by

$$\langle\phi^2\rangle = \int \phi^2\exp\left(-\frac{\gamma g}{lD}\sin^2\phi\right)d\phi \qquad (50)$$

and depends on the pendulum frequency $\omega^2 = g/l$, and, as in Eq. (48), on the ratio of the damping coefficient γ to the noise strength D. All stationary characteristics of the pendulum are defined by the distribution function (48).

1.8 Pendulum with additive noise

For non-zero temperature, the pendulum equation with damping contains the additive random noise $\eta(t)$,

$$\frac{d^2\phi}{dt^2} + \gamma\frac{d\phi}{dt} + \omega^2\sin\phi = \eta(t) \qquad (51)$$

According to the fluctuation-dissipation theorem for a stationary state, a gain of energy entering the system is exactly compensated by the energy loss to the reservoir, which gives

$$\langle \eta(0)\eta(t) \rangle = 2\gamma\kappa T\delta(t) \tag{52}$$

where κ is the Boltzmann constant. The velocity-velocity auto-correlation function gives the frequency-dependent mobility $\mu(\omega, T)$,

$$\mu(\omega, T) = \frac{1}{\kappa T} \int_0^\infty < \frac{d\phi}{dt}(0)\frac{d\phi}{dt}(t) > \exp(i\omega t)dt \tag{53}$$

Like other pendulum equations, Eq. (51) allows only numerical solutions. However, for special cases, one can obtain analytical result for $\mu(0, T)$. For the non-damped case ($\gamma = 0$) [29],

$$\mu(0, T) = \frac{\pi}{(2\pi\kappa T)^{1/2}} \frac{I_0(v) + I_1(v)}{I_0^2(v)} \exp(-v) \tag{54}$$

Here $v = \Delta E/\kappa T$, where ΔE is the barrier height. In the limit of large damping, neglecting the second derivative in (51) yields [30]

$$\mu(0, T) = \frac{1}{\gamma I_0^2(v)} \tag{55}$$

1.9 Pendulum with multiplicative noise

Compared with an oscillator equation, a pendulum equation is non-linear. The supplementary complication may be connected with additional multiplicative noise which leads to the following differential equation

$$\frac{d^2\phi}{dt^2} + [\omega^2 + \xi(t)]\sin\phi = 0 \tag{56}$$

which can be written as

$$\frac{d\phi}{dt} = \Omega; \quad \frac{d\Omega}{dt} = -[\omega^2 + \xi(t)]\sin\phi \tag{57}$$

The fluctuation of frequency can be interpreted physically as a random vibrating suspension axis. The interplay of noise with non-linearity gives rise to a variety of phenomena. The analysis of Eq. (56) is quite different for white noise and for color noise [31]. The Fokker-Planck equation associated with the Langevin equation

(56) with white noise has the following form,

$$\frac{\partial P}{\partial t} = -\frac{\partial}{\partial \phi}(\Omega P) + \frac{\partial}{\partial \Omega}[\omega^2 P \sin \phi] + \frac{D}{2}\sin^2 \phi \frac{\partial^2 P}{\partial \Omega^2} \quad (58)$$

According to the first equation in (57), the angle ϕ varies rapidly compared with Ω. Therefore, in the long-time limit, one can assume [32] that the angle ϕ is uniformly distributed over $(0, 2\pi)$. Hence, one can average Eq. (58) over ϕ, which gives a Gaussian distribution for the marginal distribution $P_1(\Omega)$,

$$P_1(\Omega) = \frac{1}{(\pi Dt)^{1/2}} \exp\left(-\frac{\Omega^2}{Dt}\right) \quad (59)$$

Accordingly, one obtains for the energy $E \approx \Omega^2/2$,

$$P_1(E) = \left(\frac{2}{\pi Dt}\right)^{1/2} E^{-1/2} \exp\left(-\frac{2E}{Dt}\right) \quad (60)$$

From (60), it follows that $\langle E \rangle = Dt/4$.

For color noise, one cannot write the exact Fokker-Planck equation. However, if one assumes a power-law dependence of ϕ as a function of time, the self-consistent estimate gives $\langle E \rangle \approx t^{1/2}$. These qualitative arguments are confirmed by more rigorous analysis and by numerical calculations [31].

1.10 Pendulum with multiplicative and additive sources of noise

In the two previous sections, we considered a pendulum with additive and multiplicative noise, respectively. Sometimes, however, the situation is different, as in the case of a stochastically vibrating pivot, which is driven stochastically by both multiplicative and additive noise. Such process is described by the following Langevin equations

$$\frac{d\phi}{dt} = \Omega; \quad \frac{d\Omega}{dt} = -\omega^2 \sin \phi - 2\gamma \Omega + \frac{\sqrt{\mu}}{l}\xi(t)\sin \phi + \sqrt{\alpha}\eta(t)$$

$$(61)$$

The two sources of noise are both assumed to be white noise with zero mean and delta correlation

$$\langle \xi(t) \rangle = \langle \phi(t) \rangle = 0; \quad \langle \xi(t_1)\eta(t_2) \rangle = 2A\delta(t_1 - t_2) \tag{62}$$

The Langevin equations (61) is equivalent [5] to the following Fokker-Planck equation for the distribution function $P(\phi, \Omega, t)$,

$$\frac{\partial P}{\partial t} = -\Omega \frac{\partial P}{\partial \phi} + \omega^2 \sin\phi \frac{\partial P}{\partial \Omega} + 2\gamma \frac{\partial}{\partial \Omega}(\Omega P) + \left(\alpha + \frac{\mu}{l^2}\sin^2\phi\right)\frac{\partial^2 P}{\partial \Omega^2} \tag{63}$$

After introducing the dimensionless variables $t_1 = \omega t$, $\Omega_1 = \Omega/\omega$, $P(\phi, \Omega, t) = P_1(\phi, \Omega_1, t_1)$ [33], the dimensionless Fokker-Planck equation takes the following form,

$$\frac{\partial P_1}{\partial t_1} = -\Omega_1 \frac{\partial P_1}{\partial \phi} + \sin\phi \frac{\partial P_1}{\partial \Omega_1} + 2G\frac{\partial}{\partial \Omega_1}(\Omega_1 P_1) + (\varepsilon + \Delta \sin^2\phi)\frac{\partial^2 P}{\partial \Omega_1^2} \tag{64}$$

where $G = \gamma/\omega^2$, $\varepsilon = \alpha/\omega^3$ and $\Delta = \mu/l^2\omega^3$. One assumes that after some time t_2 the stochastic system approaches a smooth steady state $P_1(\phi, \Omega_1, t_2)$, which is the solution of Eq. (64) with $\partial P_1/\partial t_1 = 0$. The point (ϕ, Ω_1), which is a stable point of the system, is the local maximum of the stationary probability distribution $P_1(\phi, \Omega_1; t_2)$. For a function $P_1(\phi, \Omega_1; t_2)$ of two variables ϕ and Ω_1, the necessary and sufficient conditions of the upper equilibrium point ($\phi = \pi$, $\Omega_1 = 0$) to be a stable point is given by the following condition [33]

$$\frac{\partial^2}{\partial \Omega^2}P_1(\pi, 0; t_2)\frac{\partial^2}{\partial \phi^2}P_1(\pi, 0; t_2) - \frac{\partial^2}{\partial \phi \partial \Omega}P_1(\pi, 0; t_2) > 0 \tag{65}$$

As is shown in [33], for weak multiplicative noise ($A \ll G$ and $A \ll 1$), there is no stabilization of the system. However, the numerical calculations show that for strong multiplicative noise, the system becomes stable at the upper equilibrium point $\phi = \pi$. This stability cannot be predicted by the WKB theory, which assumes weak multiplicative noise.

The quantity which characterizes the dynamics is the average velocity $\langle d\phi/dt \rangle$,

$$\left\langle \frac{d\phi}{dt} \right\rangle = \lim_{t \to \infty} \frac{\phi(t)}{t}; \tag{66}$$

Due to its nonlinearity, Eq. (61) allows only numerical solutions. We briefly describe three cases, where both sources of noise are white or one is dichotomous and the influence of correlation between the sources of noise.

1.10.1 *Two sources of white noise*

The Fokker-Planck equation (Stratonovich interpretation), which corresponds to Eq. (61) with white additive and multiplicative noise, has the following form [5]

$$\frac{\partial P}{\partial t} = -\frac{\partial}{\partial \phi}[b - (\omega^2 + D_2 \cos \phi) \sin \phi]P + \frac{\partial^2}{\partial \phi^2}[(D_1 + D_2 \sin^2 \phi)]P \tag{67}$$

where D_1 and D_2 are the strength of the multiplicative and the additive noise, respectively.

For the stationary state, $\partial P/\partial t = 0$, and one obtains [34]

$$P_{st} \approx C \left[\left(1 + \frac{D_2}{D_1} \sin^2 \phi \right)^{1/2} + \frac{1}{D_1} \left(b + \omega^2 \sin \phi \right) \right]^{-1} \tag{68}$$

As one can see from Eq. (68), an increase of multiplicative noise leads to the increase and narrowing of the peaks of P_{st}, whereas the increase of additive noise leads to their decrease and broadening.

For $D_2 < D_1$ (weak additive and no multiplicative noise), the average velocity $\langle d\phi/dt \rangle$ is given by

$$\left\langle \frac{d\phi}{dt} \right\rangle_{D_2 < D_1} = \sqrt{\omega^4 - b^2} \exp\left(\frac{\pi b}{D_1} \right)$$

$$\times \exp\left[-\frac{2\sqrt{\omega^4 - b^2}}{D_1} - \frac{2b}{D_1} \sin^{-1}\left(\frac{b}{\omega^2} \right) \right] \tag{69}$$

whereas for weak noise with $D_2 > D_1$ (low additive noise),

$$\left\langle \frac{d\phi}{dt} \right\rangle_{D_2 > D_1} = \sqrt{\omega^4 - b^2} \exp\left(-\frac{\pi b}{\sqrt{D_1 D_2}}\right) \left(\frac{\omega^2 - \sqrt{\omega^4 - b^2}}{\omega^2 + \sqrt{\omega^4 - b^2}}\right)^{\omega^2/D_2}$$
(70)

Comparison of (69) and (70) shows that adding multiplicative noise leads to an increase in the angular velocity in a system subject only to weak additive noise.

In the opposite limiting case of strong additive noise, $D_1 \to \infty$, Eq. (69) reduces to the following form,

$$\left\langle \frac{d\phi}{dt} \right\rangle_{D_1 \to \infty} = \frac{b\pi^2}{D_1} \left(1 + \frac{D_2}{D_1}\right)^{-1/2} \left[\int_0^\pi dz(D_1 + D_2 \sin^2 z)\right]^{-2}$$
(71)

One concludes that in the presence of one sort of noise, $\langle d\phi/dt \rangle$ is larger for additive noise if the strength of the noise is small, whereas for strong noise, multiplicative noise is more effective. If both types of noise are present, $\langle d\phi/dt \rangle$ increases in the presence of strong multiplicative noise.

1.10.2 *Multiplicative dichotomous and additive white noise*

If the additive noise is white with strength $2D$, but the multiplicative noise is dichotomous with two states $\pm\sigma$ and the mean waiting time is $\lambda/2$ in each of two states, it is convenient to replace the Fokker-Planck equation (67) by two following equations for $P_+(\phi, t)$ and $P_-(\phi, t)$, which correspond to the evolution of $\phi(t)$ subject to noise of strength σ and $-\sigma$, respectively,

$$\frac{\partial P_+}{\partial t} = -\lambda(P_+ - P_-) - \frac{\partial}{\partial \phi}\left\{[b - (\omega^2 + \sigma)\sin\phi]P_+ - D\frac{\partial P_+}{\partial \phi}\right\}$$

$$\frac{\partial P_-}{\partial t} = \lambda(P_+ - P_-) - \frac{\partial}{\partial \phi}\left\{[b - (\omega^2 - \sigma)\sin\phi]P_- - D\frac{\partial P_-}{\partial \phi}\right\}$$
(72)

In the limit of $\sigma \to \infty$ and $\lambda \to 0$ with $\sigma^2 \cdot \lambda = D$, we recover the results obtained in the previous section for two souces of white noise.

For $b = 0$, replacing equations (72) by the equation for $P = P_+ + P_-$ yields the stationary value of the probability density,

$$P_{st}(\phi) = C \frac{1}{\sqrt{1 + \kappa^2 \sin \phi}} \left[\frac{\sqrt{1 + \kappa} + \sqrt{\kappa} \sin \phi}{\sqrt{1 + \kappa} - \sqrt{\kappa} \sin \phi} \right]^{\omega^2/2D\sigma\sqrt{1+\kappa}} \tag{73}$$

where $\kappa = D^2/\sigma^2$ and C is the normalization factor. When the strength of the multiplicative noise is small, $D < D_{cr}$, $P_{st}(\phi)$ has a simple maximum at $\phi = 0$. For $D > D_{cr}$, $P_{st}(\phi)$ has double maxima at $\phi = 0$ and $\phi = \pi$ and a minimum at $\phi = \cos^{-1}(-\omega^2/D)$. The detailed numerical analysis gives the following results [35]:

1. Two sources of dichotomous noise are able to produce a flux, whereas each by itself is unable to produce flux in this region of bias. This effect also occurs for two sources of white noise.
2. The simultaneous action of two sources of noise can be larger than each source by itself in some region of the averaged flux-bias plane, but smaller in other regions.

1.10.3 *Correlated additive and multiplicative noise*

Thus far we have considered the additive and multiplicative noise as being independent. Correlations between different sources of noise may occur, for example, when strong external noise leads to an appreciable change in the external structure of the system and hence in its internal noise [36]. Consider an overdamped pendulum with correlated sources of additive and multiplicative noise, which describes overdamped Brownian motion in the periodic potential,

$$\frac{d\phi}{dt} + [b + \xi(t)] \sin \phi = [a + \eta(t)] \tag{74}$$

where $\xi(t)$ and $\eta(t)$ are zero mean noise with correlation functions

$$\langle \xi(t_1)\xi(t_2) \rangle = 2D\delta(t_2 - t_1); \quad \langle \eta(t_1)\eta(t_2) \rangle = 2\alpha\delta(t_2 - t_1);$$

$$\langle \xi(t_1)\eta(t_2) \rangle = 2\lambda\sqrt{D\alpha}\delta(t_2 - t_1); \tag{75}$$

The Fokker-Planck equation for the distribution function $P(\phi, \Omega, t)$ associated with the Stratonovich type Langevin equation (74) has the following form [37]

$$\frac{\partial P}{\partial t} = -\frac{\partial}{\partial \phi}[A(\phi)P] - \frac{\partial^2}{\partial \phi^2}[B(\phi)P] \tag{76}$$

with

$$A(\phi) = [a - b \sin \phi] - [\lambda\sqrt{D\alpha} - D \sin \phi] \cos \phi;$$
$$B(\phi) = D \sin^2 \phi - 2\lambda\sqrt{D\alpha} \sin \phi + \alpha \tag{77}$$

By introducing the generalized potential

$$\Phi(\phi) = -\int_0^\phi \frac{A(x)}{B(x)} dx \tag{78}$$

one obtains from Eq. (76) the stationary current J as [5]

$$J = \left\{ \int_0^{2\pi} dx \left[\int_x^{x+dx} dy \exp[\Phi(y)]\{B(x) \exp[\Phi(x)] \right. \right.$$
$$\left. \left. - B(x + 2\pi) \exp[\Phi(x + 2\pi)]\}^{-1} \right] \right\}^{-1} \tag{79}$$

Substituting (77) into (79) one obtains the general potential $\Phi(\phi)$ for $|\lambda| < 1$. For uncorrelated noise $\lambda = 0$ and

$$\Phi(\phi) = -\int_0^\phi \frac{[a - b \sin y] + D \sin y \cos y}{D \sin^2 y + \alpha}; \quad B(y) = D \sin^2 y + \alpha \tag{80}$$

For the case of completely correlated noise, $\lambda = \pm 1$, Eq. (78) gives

$$\Phi(\phi) = -\int_0^\phi \int_0^\phi \frac{[a - b \sin y] \pm \sqrt{\alpha D} \cos y + D \sin y \cos y}{D \sin^2 y - \lambda\sqrt{D\alpha} + \alpha};$$
$$B(y) = D \sin^2 y \pm \sqrt{D\alpha} + \alpha \tag{81}$$

In this manner we obtained equation (79) for the steady state current J as a function of $\Phi(x)$ and $B(y)$, which depend on the noise strengths α and D for different intensity of the correlations of noise λ

in Eqs. (77)–(81). The intensive numerical computations [36] lead to
the following conclusions:

(1) For uncorrelated noise, $\lambda = 0$, the direction of current does
not change with increasing intensities of noise α and D or their
ratio α/D. The current decreases with increasing additive noise
strength α and α/D, and increases with increasing multiplicative noise D.

(2) New phenomena occur for $-1 \leq \lambda < 0$ and $0 < \lambda \leq 1$, namely,

(2a) Current reversal. For negative λ, $-1 \leq \lambda < 0$, and positive λ,
$0 < \lambda \leq 1$, the direction of the current reverses with increasing
of α/D.

(2b) Existence of extremum. With increasing of α/D, for positive
noise correlation λ, $0 < \lambda \leq 1$, the direction of the current
changes from negative to positive value, reaching a minimum,
whereas for negative λ, $-1 \leq \lambda < 0$, the change is from positive to negative value possessing a maximum. For the case of
completely correlated noise, $\lambda = \pm 1$, one can have or minimum
or maximum.

(2c) Symmetric current. For some values of the parameters, the current is symmetric. The position of the point of the reverse
depends on the intensity of the noise correlations λ and the
ratio of the noise parameters α/D.

We conclude that in the presence of correlation between sources
of noises ($\lambda \neq 0$), the direction of the currents changes with a change
of the noise strengths. We shall restrict our consideration by the
simplest case of two correlated additive and multiplicative noises $\xi(t)$
and $\eta(t)$ in Eq. (61) with the same type (75) of correlation.

A simple calculation gives [38]

$$\left\langle \frac{d\phi}{dt} \right\rangle = \left\{ \int_0^{2\pi} \left[\frac{d\phi_1}{B(\phi_1)\exp[\Psi(\phi_1)] - B(\phi_1 + 2\pi)\exp[\Psi(\phi_1 + 2\pi)]} \right] \right.$$
$$\left. * \int_{\phi_1}^{\phi_1 + 2\pi} d\phi_2 \exp[\Psi(\phi_2)] \right\}^{-1} \tag{82}$$

where

$$\Psi(z) = -\int_0^z dx \, A(x)B(x);$$

$$B(x) = \alpha \sin^2 x - 2\lambda\sqrt{D\alpha} \sin x + D$$

$$A(x) = b - \omega^2 \sin x - \lambda\sqrt{D\alpha} \cos x + \alpha \sin x \cos x \tag{83}$$

Extensive numerical calculations has been performed on these equation for $-1 \le \lambda \le 1$, showing the following results,

1. For non-zero values of λ, the direction of $d\phi/dt$ reverses when the ratio D/α decreases.
2. As D/α increases, $\langle d\phi/dt \rangle$ possesses a minimum changing from negative to positive values for $\lambda > 0$, and a maximum changing from positive to negative value for $\lambda < 0$. Both the maximum and minimum exist for completely correlated noise, $\lambda = 1$.

Chapter 2

Oscillator with a Random Mass

A harmonic oscillator is a simple system in which the force is proportional to the displacement x of mass m from the equilibrium position $x = 0$ and pointing in the opposite direction. Taking friction into account, assumed to be proportional to the velocity dx/dt, leads to the following equation of motion

$$m\frac{d^2x}{dt^2} + \gamma\frac{dx}{dt} + kx = 0 \tag{84}$$

This model has been applied everywhere, from quarks to cosmology. Moreover, a person who is worried by oscillations in the stock market can relax to classical music produced by the oscillations of stringed instruments. The ancient Greeks already had a general idea of oscillations and used them in musical instruments. Regarding practical applications, we note the Galilean discovery of the universality of the period for small oscillations, which was used in 1602 for measuring the human pulse. Many other applications have been found in the last 400 years.

Equation (84) describes a pure mechanical system (zero temperature) [28]. Here we consider the stochastic generalization of this equation. For non-zero temperatures, the deterministic equation (84) has to be supplemented by thermal noise $\eta(t)$,

$$m\frac{d^2x}{dt^2} + \gamma\frac{dx}{dt} + kx = \eta(t) \tag{85}$$

Equation (85) with $k = 0$ describes Brownian motion, where the force acting on the Brownian particle consists of the systematic force $-\gamma(dx/dt)$ and the random force $\eta(t)$. An additive random force, due to the random number of molecules of the surrounding medium that collide with the Brownian particle from opposite sides, results in random zigzag motion. There are many books describing different aspects and many applications of Brownian motion [39].

The random force $\eta(t)$ enters equation (85) additively. When the noise has an external origin rather than an internal origin, the associated noise enters the equation of motion multiplicatively. If the noise arises from the fluctuations of the potential energy, $U = kx^2[1 + \xi(t)]/2$, the equation of motion (84) will take the following form

$$m\frac{d^2x}{dt^2} + \gamma\frac{dx}{dt} + k[1 + \xi(t)]x = 0 \qquad (86)$$

Another possibility for the generalization of the dynamic equation (84) is the incorporation of random damping

$$m\frac{d^2x}{dt^2} + \gamma[1 + \xi(t)]\frac{dx}{dt} + kx = 0 \qquad (87)$$

The many applications Eq. (86), in addition to those mentioned earlier, include wave propagation in a random medium [40], spin precession in a random external field [41], turbulent flow on the ocean surface [42], as well as biology (population dynamics [43]).

For the stochastic oscillator with additive noise described by Eq. (85), one can perform the following replacement

$$\gamma \to m\gamma; \quad k \to mk; \quad \eta \to m\eta \qquad (88)$$

which allows one to set $m = 1$ by the appropriate choice of units. Then the stability conditions do not depend on the mass of a particle. However, the situation is different when the appropriate equations, in addition to additive noise, contain the external fluctuations, which are described by multiplicative noise (Eq. (86) for random frequency and Eq. (87) for random damping). In the latter cases, one cannot eliminate the mass from the equation of motion, which leads to new phenomena. In particular, a new type of Brownian motion appears, when the restrictions appear not only on the size of the Brownian particle (larger than the size of the surrounding molecules) but also on its mass [27]. As distinct from [44] where the Fokker-Planck equation has been used, we here use the Langevin equation for the analysis of the asymptotic value of the second moment averaged over noise, which defines the energetic (mean-square) instability.

Recently [28], we considered another way of introducing multiplicative noise, namely, via a fluctuating mass term,

$$[1 + \xi(t)]m\frac{d^2x}{dt^2} + 2\gamma\frac{dx}{dt} + \omega^2 x = 0 \tag{89}$$

Upon multiplying the last equation by $1 - \xi(t)$, one obtains

$$(1 - \xi^2)m\frac{d^2x}{dt^2} + (1 - \xi(t))\left(2\gamma\frac{d}{dt} + \omega^2\right)x = 0 \tag{90}$$

Since the oscillator mass is positive, the condition $\xi^2 < 1$ must be satisfied. For symmetric dichotomous noise $\xi = \pm\sigma$, this condition reduces to $\sigma^2 < 1$. For the asymptotic case of small oscillations of the mass, $\sigma^2 \ll 1$, Eq. (90) can be rewritten as

$$m\frac{d^2x}{dt^2} + \left(2\gamma\frac{d}{dt} + \omega^2\right)x = \xi(t)\left(2\gamma\frac{d}{dt} + \omega^2\right)x \tag{91}$$

Therefore, small dichotomous fluctuations of mass are equivalent to simultaneous fluctuations of the frequency and the damping coefficient.

The simplest form of noise $\xi(t)$ is white noise with the correlators $\langle\xi(t_1)\xi(t_2)\rangle = \alpha\delta(|t_1 - t_2|)$. It is clear that for our problem, the noise $\xi(t)$ in Eq. (89) cannot be white since large negative noise, $\xi(t) \ll 0$, implies a negative mass of the oscillator. The last condition requires some additional constraint on the random force. Another possibility is to consider the positive random force $\xi^2(t)$, which represents the fact that the mass of the oscillator may only increase due to the adhesion of the molecules of the surrounded media,

$$m[1 + \xi^2(t)]\frac{d^2x}{dt^2} + \kappa\frac{dx}{dt} + kx = m\eta(t), \tag{92}$$

which can be rewritten as

$$[1 + \xi^2(t)]\frac{d^2x}{dt^2} + \gamma\frac{dx}{dt} + \omega^2 x = \eta(t) \tag{93}$$

where $\gamma = \kappa/m$ is the damping coefficient and $\omega = \sqrt{k/m}$ is the frequency.

The quadratic noise $\xi^2(t)$ can be written as

$$\xi^2 = \sigma^2 + \Delta\xi \tag{94}$$

where $\sigma^2 = AB$ and $\Delta = A - B$. For $\xi = A$, one obtains $\xi^2 = AB + (A - B)A = A^2$, and for $\xi = -B$, one obtains $\xi^2 = AB - (A - B)B = B^2$. Therefore Eq. (93) takes the following form

$$[1 + \sigma^2 + \Delta\xi]\frac{d^2x}{dt^2} + \gamma\frac{dx}{dt} + \omega^2 x = \eta(t) \tag{95}$$

This model was originally proposed in the context of a Brownian particle undergoing random adsorption and desorption of surrounded particles. A multiplicative random force arises from the adhesion of surrounding molecules which stick to the Brownian particle for some (random) time, thereby changing its mass. Usual Brownian motion does not involve the restoring force $\omega^2 x$, but we include this term in (95) to include the more general problem of a stochastic harmonic oscillator. It turns out that usual condition for the Brownian motion — the small size of the surrounding particles compared with the size of a Brownian particle — has to be supplemented by an additional requirement on the mass of Brownian particle for the system to be stable.

There are many situations in chemical and biological solutions in which the surrounding medium contains molecules which are capable of both colliding with the Brownian particle and adhering to it for a random time (see, for example, the Brownian motion of the Hyaluronan molecule [45]). There are also applications of a variable-mass oscillator [46]. Modern applications of such a model include a nanomechanical resonator, which randomly absorbs and desorbs molecules [47]. The diffusion of clusters with randomly growing masses has also been considered [48]. There are many other applications of an oscillator with a random mass [49], including ion-ion reactions [50, 51], electrodeposition [52], granular flow [53], cosmology [54, 55], film deposition [56], traffic jams [57, 58], and the stock market [59, 60].

2.1 White and colored noise

In the following we will consider noise $\xi(t)$ with $\langle\xi(t)\rangle = 0$ having the correlator

$$\langle\xi(t_1)\xi(t_2)\rangle = r(|t_1 - t_2|) \tag{96}$$

Two integrals characterize the fluctuations: the strength of the noise D,

$$D = \int_0^\infty \langle\xi(t)\xi(t + z)\rangle dz \tag{97}$$

and the inverse correlation time λ^{-1},

$$\lambda^{-1} = \frac{1}{D}\int_0^\infty \langle\xi(t)\xi(t + z)\rangle z dz \tag{98}$$

2.1.1 *White noise*

Traditionally one considers two different forms of noise, white and colored noise. For white noise, the function $r|t_1 - t_2|$ has the form of a delta-function,

$$\langle\xi(t_1)\xi(t_2)\rangle = D\delta(t_1 - t_2) \tag{99}$$

The name "white" noise comes from the fact that the Fourier transform of (99) is "white", being constant without any characteristic frequency. Equation (99) means that $\xi(t_1)$ and $\xi(t_2)$ are statistically independent, no matter how close t_1 and t_2 are. This extreme assumption, which leads to a non-physical infinite value of $\langle\xi^2(t)\rangle$ in (99), means that the correlation time τ is not zero, as assumed in (99), but smaller than all other characteristic times in the problem. It is clear that for our problem, the noise in (89) cannot be white since a large negative noise, $\xi(t) \ll 0$, implies a negative mass of the oscillator.

2.1.2 *Colored noise*

All non-white sources of noise are called colored noise. We consider a special type of noise, the so-called dichotomous noise (random telegraph signal), which randomly jumps between two different values, either $\pm\sigma$ (symmetric dichotomous noise) or A and $-B$ (asymmetric dichotomous noise), which are characterized by the Ornstein-Uhlenbeck correlation function. For the symmetric noise, the correlation function has the following form

$$\langle \xi(t_1)\xi(t_2) \rangle = \sigma^2 \lambda \exp\left[-\lambda|t_1 - t_2|\right] \qquad (100)$$

White noise (99) is defined by its strength D while the Ornstein-Uhlenbeck noise is characterized by two parameters, σ^2 and λ. The transition from the Ornstein-Uhlenbeck noise (100) to white noise (99) occurs in the limit $\sigma^2 \to \infty$ and $\lambda \to 0$, with $\sigma^2\lambda = D$ in (100).

The "energetic" instability corresponds to a negative second moment $\langle x^2 \rangle$. For additive white noise and white or colored multiplicative noises, we found the values of the particle mass for which the second moment is positive. We have restricted ourselves to the analysis of this type of instability, although the more precise (and cumbersome) method involves the analysis of the Lyapunov indices [40], which we shall consider in Section 2.8.

Recently [44], we explored some other features of the problem which had previously escaped notice. Primary among them is the fact that the statistics of the x, v and E are anomalous with power-law tails. Thus, even for parameters for which $\langle E \rangle$ is finite, higher moments of E diverge exponentially in time. This behavior is also reflected in a single time trace, with the dynamics having a burst of intermittent character. Thus, there are an infinite series of transitions in the model, corresponding to the onset of convergence of higher and higher moments.

2.1.3 *Brownian motion with adhesion*

Comparing Eqs. (89) and (95) shows that the solution of Eq. (89) can be obtained from the solution of Eq. (95) by the replacement $1 + \sigma^2$ and Δ by unity. Therefore, we will consider the stochastic equation (95) in the external field $A\cos(\omega t)$, which can be rewritten as an equivalent system of two first-order differential equations,

$$\frac{dx}{dt} = v; \quad \frac{dv}{dt} = -\sigma^2 \frac{dv}{dt} - \Delta \xi \frac{dv}{dt} - 2\gamma v + \eta(t) + A \cos(\omega t) \quad (101)$$

Averaging Eqs. (101) gives

$$\frac{d\langle x \rangle}{dt} = \langle v \rangle;$$

$$(1 + \sigma^2) \frac{d\langle v \rangle}{dt} + \Delta \left(\frac{d}{dt} + \lambda \right) \langle \xi v \rangle + 2\gamma \langle v \rangle = A \cos(\omega t) \quad (102)$$

For the stationary state $(d/dt \ldots = 0)$, one gets from Eqs. (102), $\langle v \rangle = 0$; $\langle \xi v \rangle = (A/\Delta\lambda) \cos(\omega t)$ and from Eqs. (101), (102) and (99), one obtains,

$$\langle \eta v \rangle = \frac{D}{2\gamma + (1 + \sigma^2)\lambda} - \frac{\Delta\lambda}{2\gamma + (1 + \sigma^2)\lambda}$$

$$\langle \xi \eta v \rangle = \frac{D}{2\gamma + (1 + \sigma^2)\lambda}$$

$$\langle \eta x \rangle = \frac{D}{\lambda[2\gamma + (1 + \sigma^2)\lambda]} \quad (103)$$

In the last equation we have used the simplest splitting of the correlators,

$$\langle \xi \eta v \rangle = \langle \xi \eta \rangle \langle v \rangle = 0 \quad \langle \xi \eta x \rangle = \langle \xi \eta \rangle \langle x \rangle = D \quad (104)$$

Additional relations between averaged values can be obtained by multiplying the first equation in (101) by $2x$ and the second by $2v$, and averaging the resulting equations,

$$\frac{d}{dt} \langle x^2 \rangle = 2\langle xv \rangle$$

$$\frac{d}{dt} \langle v^2 \rangle = -\sigma^2 \frac{d}{dt} \langle v^2 \rangle - \Delta \left(\frac{d}{dt} + \lambda \right) \langle \xi v^2 \rangle - 4\gamma \langle v^2 \rangle + 2\langle v\eta \rangle$$

$$(105)$$

Multiplying Eqs. (101) by v and x, respectively, and summing these equations gives

$$\frac{d}{dt} (xv) = (1 + \sigma^2) v^2 - \sigma^2 \frac{d}{dt} (xv) - \Delta \xi x \frac{dv}{dt} - 2\gamma xv$$

$$+ x\eta(t) + xA \cos(\omega t) \quad (106)$$

Averaging Eq. (106) results in

$$\frac{d}{dt}\langle xv \rangle = (1 + \sigma^2)\langle v^2 \rangle - \Delta\left(\frac{d}{dt} + \lambda\right)\langle \xi xv \rangle$$

$$+ \Delta\langle \xi v^2 \rangle - \left(\sigma^2\frac{d}{dt} + 2\gamma\right)\langle xv \rangle$$

$$+ \frac{D}{\lambda[2\gamma + (1 + \sigma^2)\lambda]} + \langle x \rangle A\cos(\omega t) \qquad (107)$$

For the stationary state, Eqs. (105) and (107) can be reduced to the following form,

$$\langle xv \rangle = 0;$$

$$\Delta\lambda\langle \xi v^2 \rangle + 4\gamma\langle v^2 \rangle = \frac{2D}{[2\gamma + (1 + \sigma^2)\lambda]}(1 + \sigma^2)\langle v^2 \rangle - \Delta\lambda\langle \xi xv \rangle$$

$$+ \Delta\langle \xi v^2 \rangle + \frac{D}{\lambda[2\gamma + (1 + \sigma^2)\lambda]}$$

$$+ \langle x \rangle A\cos(\omega t) = 0 \qquad (108)$$

Multiplying Eqs. (101) and (106) by $2x\xi$, $2\xi v$ and ξ, respectively, and averaging, one obtains for the stationary state,

$$\lambda\langle \xi x^2 \rangle = 2\langle \xi xv \rangle;$$

$$[4\gamma + \lambda(1 + \sigma^2)]\langle \xi v^2 \rangle + \Delta\sigma^2\langle v^2 \rangle - 2\langle \xi v \rangle A\cos(\omega t) = 2\langle v\xi\eta \rangle = 0$$

$$[2\gamma + (1 + \sigma^2)\lambda]\langle \xi xv \rangle - (1 + \sigma^2)\langle \xi v^2 \rangle - \Delta\sigma^2\langle v^2 \rangle$$

$$- \Delta\left\langle \xi^2\frac{d}{dt}(xv) \right\rangle + \langle \xi x \rangle A\cos(\omega t) = \langle x\xi\eta \rangle = D \qquad (109)$$

Equations (108) and (109) yield

$$\langle v^2 \rangle\left[4\gamma - \frac{\lambda\Delta^2\sigma^2}{4\gamma + \lambda(1 + \sigma^2)}\right] = \frac{2D}{2\gamma + (1 + \sigma^2)\lambda} - \frac{A^2}{4\gamma + \lambda(1 + \sigma^2)} \qquad (110)$$

For the long time behavior, the additional average has been performed, $\langle \cos^2(\omega t) \rangle = 1/2$.

For linear noise, Eq. (110) becomes

$$\frac{2D}{\lambda(\lambda + 2\gamma)} - \frac{A^2}{\lambda\left[4\gamma + \lambda\right]} = \langle v^2\rangle \left[\frac{4\gamma}{\lambda} - \frac{\sigma^2}{4\gamma + \lambda}\right] \qquad (111)$$

For $A = 0$ and $\sigma^2 = 0$, this equation reduces to

$$\langle v^2\rangle = \frac{\lambda}{4\gamma}\frac{2D}{\lambda(\lambda + 2\gamma)} = \frac{D}{2\gamma(\lambda + 2\gamma)} \qquad (112)$$

and coincides with the result for the usual Brownian motion, as required.

In summary, for the Brownian motion with adhesion in the absence of an external field, $A = 0$, the second moment $\langle v^2\rangle$ is equal to

$$\langle v^2\rangle \left[4\gamma - \frac{\lambda\Delta^2\sigma^2}{4\gamma + \lambda(1 + \sigma^2)}\right] = \frac{2D}{2\gamma + (1 + \sigma^2)\lambda} \qquad (113)$$

i.e., $\langle v^2\rangle$ becomes negative for large strength σ^2 of the mass fluctuations showing instability of the system, i.e., the system cannot reach a stationary state. In the presence of an external field, the large amplitude of this field might be the alternative source of the instability. These results are attributable to the violation of the energetic balance (fluctuation-dissipation theorem) for the usual Brownian motion. Note that a system (101) becomes stable if both the strength of the mass fluctuations and the amplitude of an external field are sufficiently large, which restores the energetic balance.

We expect that both the model of a Brownian particle with a fluctuating mass, as well as the model previously considered of an oscillator with random mass [28], will find many applications in modern science.

2.1.4 *Overdamped harmonic oscillator*

Physical intuition suggests that the action of periodic (for instance, sinusoidal) and random (for instance, white noise) forces on a physical system act in the reverse manner, namely, the latter leads to disorder while the former works in an orderly fashion. The best known counter example against this conclusion is the phenomenon of fluctuation driven transport and the stochastic resonance [61], where the noise helps to increase a weak input signal. In addition to the stochastic resonance, there are also other phenomena showing that noise may be a source of order rather than disorder. We may mention noise-induced transitions [62], noise-induce transport [63], noise-induced pattern formation [64], noise-induced resonances [65], noise-enhanced stability [66], noise-induced hypersensitivity [67], resonance activation [68], stochastic transport in ratches [69], stochastic localization [70], self-organization and dissipative structures [71], coherent stochastic resonance [72], fluctuation barrier kinetics [73] and amplification of weak signals via on-off intermittence [74].

However, according to the second law of thermodynamics, a system naturally progresses from order to disorder and not in the opposite direction. The explanation of this apparent paradox is that noise does not transfer energy to a system, playing the role of a tuner (like the emitter in a transistor), helping the system absorb more energy from the external force [75]. Here, using the simple example of an oscillator, we give other examples of such "unnatural" behavior of noise and a periodic signal. This example is complementary to that of a classical rotor, where it was shown [76] that "order and chaos are complementary rather than contradictory".

The simple equation

$$\frac{dx}{dt} = -ax + \xi(t)x \qquad (114)$$

can be solved both in the absence and in the presence of white noise $\xi(t)$ of strength D. In the former case, the solution is $x(t) = x(0)\exp(-at)$, which vanishes at $t \to \infty$. In the presence of noise, the average moment

$$\langle x(t) \rangle = x(0)\exp[(-a + D)t] \qquad (115)$$

diverges as $t \to \infty$, for $D > a$, i.e., noise plays its usual "destructive" role.

With an additional periodic force,

$$\frac{dx}{dt} = -ax + \xi(t)x + A\cos(\Omega t) \qquad (116)$$

the solution of Eq. (116) performs oscillations with frequency Ω for $a = D$, as expected for a periodic external force. However, as we will show, the situation is not so simple for nonlinear equations.

One can illustrate the appearance of stochastic resonance by equation (116) with dichotomous noise of strength σ and rate $\tau/2$ for transitions $\sigma \to -\sigma$ and $-\sigma \to \sigma$. For the limiting case $a = \xi = 0$, the particle executes periodic motion with an amplitude A/Ω. If there is no random force, $\xi = 0$, for $a \neq 0$, the particle moves along the parabola $U = ax^2/2$. For dichotomous noise $\pm\sigma$, the particle moves along the parabola $U = (a + \sigma)x^2/2$, then jumps at rate τ to the parabola $U = (a-\sigma)x^2/2$, etc. For $\sigma > a$, but $\sigma < (a^2+2a\lambda)^{1/2}$, these two parabolas have curvatures of opposite sign, and thus they act in opposite directions tending to increase (decrease) the displacement x of the particle. Their mutual influence is defined by noise which causes jumps between the parabolas and by a periodic force which determines the amplitude of oscillations along the parabolas. Accordingly, the amplitude of the stationary output signal has a maximum as a function of noise strength (stochastic resonance).

2.2 Birth-death process

The birth-death differential equation for positive x

$$\frac{dx}{dt} = ax - bx^2 \tag{117}$$

has an exact solution. The associated potential energy $U = -\int(ax - bx^2)dx$ is shown in Fig. 1. There are two fixed points, $x = 0$ and $x = a/b$, which are stable for $a < 0$ and $a > 0$, respectively.

If the parameter a fluctuates, $a \to a + \xi$, with white noise $\xi, \langle\xi(t_1)\xi(t_2)\rangle = D\delta(t_2 - t_1)$, Eq. (117) has two control parameters, a and D. For $a < 0$, the fixed point $x = 0$ is stable. For $a > 0$, the point $x = 0$ becomes unstable but most probable for $0 < a < D$. Finally, for $a > D$, the point $x = a/b$ becomes stable.

Let us now add a periodic force to Eq. (117),

$$\frac{dx}{dt} = ax - bx^2 + A\cos(\Omega t) \tag{118}$$

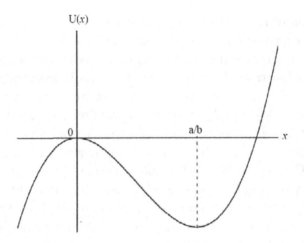

Figure 1: Single-well potential with an unstable point at $x = 0$ and a metastable point at $x = a/b$.

If the amplitude of the external field A is smaller than the barrier height, $A < a^3/6b^2$, the particle will never leave the well provided that the initial position x_0 satisfies the condition

$$x_0 < \frac{a}{2b} \left[1 - (1 + \frac{4Ab}{a^2})^{1/2} \right] \tag{119}$$

This follows from the fact that for A obeying Eq. (119), the right-hand side of Eq. (118) has two real roots, x_1 and x_2, implying that $x_1 < x < x_2$ and the particle is trapped. Although the dependence of the solution of Eq. (118) on A is physically obvious, the dependence on the field frequency Ω is not so obvious. It turns out that even a change of only 10^{-5} in the frequency changes the time at which the particle escapes from the potential well.

Consider now the joint action of both random and periodic forces,

$$\frac{dx}{dt} = ax - bx^2 + \xi(t) + A\cos(\Omega t) \tag{120}$$

The numerical solution of Eq. (120) shows that for A not too small, both the periodic force and the noise increase the escape time, i.e., these two factors act in the same direction. However, adding noise

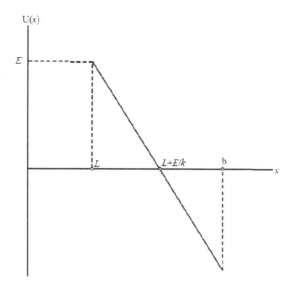

Figure 2: Triangle piece-wise potential. Reprinted figure with permission from [28]. Copyright @ 2012, World Scientific Publication Co. Pte. Ltd.

to a periodically driven system will increase the escape time for some noise strength. This effect is known as noise-enhanced stability [77].

2.3 Piece-wise potential

Since the behavior of a system is probably not too sensitive to the exact form of the nonlinear potential, we consider the simplest form of the piece-wise potential shown in Fig. 2,

$$U(x) = \left\{ \begin{array}{ll} 0 & \text{for } x < L \\ E - k(x - L) & \text{for } L < x < b \end{array} \right\} \tag{121}$$

where E is the height of the potential barrier, and b is the absorbing boundary.

The states for $0 < x < L$ are metastable and those for $L < x < b$ are unstable. The overdamped periodically driven motion of a particle in the potential (121) is described by the following equation

$$\frac{dx}{dt} = -\frac{dU}{dx} + A\sin(\Omega t) + \xi(t) \tag{122}$$

with white noise $\xi(t)$ of strength D.

Consider first the time-independent potential $(A = 0)$. If the initial position of a particle is unstable, $L < x_0 < L + E/k$, the average escape time grows in the presence of noise since the particle may jump into the potential well. For very weak noise, the probability of such jumps is very low. If only these jumps are operating, the particle will be trapped in the well for a long time (noise-enhanced stability).

Consider now $A \neq 0$. For $x_0 = 0$, the particle at $0 < x < L$ will move according to the equation $x(t) = (A/\Omega)[1 - \cos(\Omega t)]$. If $2A/\Omega < L$, the particle will always remain inside the region $(0, L)$. However, if $2A/\Omega > L$, the particle surmounts the region $(0, L)$ and its position will change with time as

$$x(t) = \frac{A}{\Omega}[1 - \cos(\Omega t)], \quad \text{for } 0 < t < t_1, \quad 0 < x(t) < L, \quad (123)$$

and

$$x(t) = k(t - t_1) + \frac{A}{\Omega}[1 - \cos(\Omega t)], \quad \text{for } t > t_1, \quad L < x < b$$

$$(124)$$

where t_1 is the time at which the particle crosses the point $x = L$,

$$t_1 = \frac{1}{\Omega} \arccos\left(1 - \frac{\Omega L}{A}\right). \quad (125)$$

Noise-enhanced stability occurs at time $t_2 = T/2 = \pi/\Omega$, (when the periodic force changes its sign), tending to return the particle to the region $(0, L)$, and the particle is still located inside the interval $(0, b)$, $x(t_2) < b$. Using (124) and (125), the latter inequality can be rewritten as

$$\frac{2A}{b} + \frac{k}{b}\left[\pi - \arccos\left(1 - \frac{\Omega L}{A}\right)\right] < \Omega \quad (126)$$

In addition to the inequality $\Omega < 2A/L$, Eq. (126) gives the conditions for the appearance of the noise-enhanced stability. Therefore, both noise and a periodic force influence the escape time of a particle in a metastable state, thereby increasing the stability of the system.

Another form of potential barrier is the rectangular bistable potential shown in Fig. 3. The barriers heights U_1 and U_2 are different

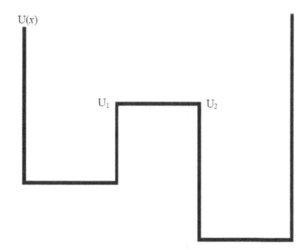

Figure 3: Simplest form of a rectangular piece-wise bistable potential. Reprinted figure with permission from [28]. Copyright @ 2012, World Scientific Publication Co. Pte. Ltd.

for the right (stable) and the left (metastable) states. For such a form of the potential, there is no force in the equation of motion, which has the following form

$$\frac{dx}{dt} = \xi(t) + A\cos(\Omega t) \tag{127}$$

For $A = 0$, one can easily solve the Fokker-Planck equation associated with the Langevin equation (127) in each of the three regions in Fig. 3 and find the integration constants from the matching conditions on two boundaries between different regions complemented by reflection boundary conditions at the walls. Finally, one can find the time-independent number of particles n_r and n_l in the right (stable) and in the left (metastable) wells, respectively. Therefore, the population of a metastable state can be increased by adding an external periodic field or by fluctuations of the barrier height. The analogous situation exists with a simple pendulum which is stable (metastable) in the vertically downward (upward) position. One can stabilize a metastable position by high-frequency harmonic vibrations of its suspension parametric oscillations of a pendulum ("Kapitza pendulum" [78]).

An external force $A \neq 0$ in Eq. (127) can be chosen as a periodic force acting on the left well or as a random force acting on the barrier. For both cases, an external force tends to equalize the populations as $t \to \infty$ (stabilizing the metastable state), and even reversing the populations of these states. Our choice of the periodic signal does not introduce an additional force into the equation of motion, and the periodic signal enters only in the matching conditions. With the help of an external periodic field, one can increase the population of the left (metastable) state, or even reverse the populations [28].

2.4 Simple treatment of correlated multiplicative and additive sources of noise

A variety of phenomena in physics, chemistry and biology are modelled by the stochastic differential equation

$$\frac{dx}{dt} = f(x) + g(x)\xi(t) + \eta(t) \tag{128}$$

which, in particular, describes the overdamped one-dimensional motion of a particle in an external field $U(x)$, where $f(x) = -dU/dx$, subject to multiplicative noise $\xi(t)$ and additive noise $\eta(t)$. An important special case of (128) is the bistable potential $U(x) = -ax^2/2 + bx^4/4$ with stochastically varying barrier curvature, i.e. where

$$f(x) = ax - bx^3; \quad g(x) = x \tag{129}$$

Additive noise arises from the fast dynamics of other (in addition to x) degrees of freedom or from the non-zero temperature of a system (thermal noise) whereas multiplicative noise is related to the stochastic nature of external fields or boundary conditions. The simplest assumption is that both $\xi(t)$ and $\eta(t)$ are Gaussian white noise with zero means and the correlation functions

$$\langle(\xi(t_1)\xi(t_2)\rangle = 2D\delta(t_1 - t_2); \quad \langle(\eta(t_1)\eta(t_2)\rangle = 2\alpha\delta(t_1 - t_2) \tag{130}$$

By use of the Stratonovich interpretation of the stochastic differential equation (128), the Fokker-Planck equation for the probability function $P(x, t)$, corresponded to (128) and (130) is given by

$$\frac{\partial P}{\partial t} = -\frac{\partial}{\partial x}[A(x)P] + \frac{\partial^2}{\partial x^2}[B(x)P] \tag{131}$$

where

$$A(x) = f(x) + \alpha g(x); \quad B(x) = D + \alpha g^2(x) \tag{132}$$

Usually one considers non-correlated additive and multiplicative noise

$$\langle \xi(t)\gamma(t) \rangle = 0 \tag{133}$$

However, there are some situations where the latter condition is violated. This happens when both sources of noise have the same origin, as in laser dynamics [79] or when strong external noise leads to an appreciable change in the internal structure of a system and hence in internal noise. The influence of the correlation between noises on the dynamics of a system allows a simple analysis [38]. We first assume that additive and multiplicative noise are delta-correlated with the parameter R measuring the strength of these correlations,

$$\langle \xi(t_1)\eta(t_2) \rangle = 2R\sqrt{\alpha D}\delta(t_1 - t_2) \tag{134}$$

Rule 1. To obtain the Fokker-Planck equation for a system with correlated sources of noise described by Eqs. (128)–(134), one has to replace $g(x)$ and D in the Fokker-Planck equation (131) by

$$g(x) \rightarrow g(x) + R\sqrt{\frac{D}{\alpha}}; \quad D \rightarrow D(1 - R^2) \tag{135}$$

(the case $R = 1$ has to be considered separately).

So far we have assumed the delta-correlations between multiplicative and additive noises. More complicated "colored" correlation is characterized by the non-zero correlation time τ. We consider the exponential Gaussian correlation of the form

$$\langle \xi(t_1)\gamma(t_2) \rangle = \frac{R\sqrt{\alpha D}}{\tau} \exp\left(\frac{|t_1 - t_2|}{\tau}\right) \tag{136}$$

It turns out [80] that the Fokker-Planck equation for correlations of the form (136) can be obtained from the Fokker-Planck equation for the non-correlated case (133) by replacing R by $R/(1 + 2\alpha\tau)$. From this result follows:

Rule 2. In order to derive the Fokker-Planck equation for a system with exponentially correlated multiplicative and additive noises, one have to replace R in the Fokker-Planck equation (131) for the delta-correlated noise by $R/(1+2\alpha\tau)$. The proof of these two Rules is given in [38].

2.5 Mass dependence instability of an oscillator with multiplicative noise

The simplest form of colored noise is dichotomous noise (random telegraph signal), defined in (100). After cumbersome calculations [28], the stationary $(d/dt\ldots = 0)$ second moment $\langle x^2 \rangle$ has the following form for random frequency

$$\langle x^2 \rangle_\omega = \alpha/m^2 \left[\frac{2\gamma\omega^2}{m^2} - \frac{4\omega^4\sigma^2/m^2(\lambda + 2\gamma/m)^2}{(\lambda + \gamma/m)[4\omega^2/m + \lambda(\lambda + 2\gamma/m)]} \right]^{-1}$$

(137)

and for random damping

$$\langle x^2 \rangle_\eta = \frac{\alpha}{m^2} \{ [2\omega^2/m + \lambda(\lambda + 2\gamma/m)](\lambda + 2\gamma/m) + 2\omega^2\lambda/m$$

$$+ 2\lambda\gamma^2\sigma^2/m^2 \} \{ 2\gamma(\omega^2/m^2)$$

$$+ [2\omega^2/m + \lambda(\lambda + \gamma/m)(\lambda + 2\gamma/m)]$$

$$- 4\gamma^2\sigma^2(\omega^2/m^3)[2\omega^2/m + \lambda(\lambda + \gamma/m)] \}^{-1}$$

(138)

For different strength of the noise, the average second moments for random frequency and random damping depend on the mass of the particle. Moreover, it follows from Eqs. (137) and (138) that these moments become negative, i.e., the systems are non-stable for the mass-dependent noise strength, $\sigma^2 > \sigma_0^2$. For random frequency and random damping, the boundary values of the noise strength, $(\sigma_0^2)_\omega$ and $(\sigma_0^2)_\eta$, have the following form

$$(\sigma_0^2)_\omega = \frac{2\gamma(m\lambda + 2\gamma)(m\lambda^2 + 4\lambda\gamma + 4k)}{k(m\lambda + 4\gamma)^2}$$

(139)

and

$$(\sigma_0^2)_\eta = \frac{8\gamma k + 4km\lambda + 6m\gamma\lambda^2 + m^2\lambda^3 + 8\gamma^2\lambda}{4\gamma(m\lambda^2 + 2k + 2\gamma\lambda)} \qquad (140)$$

Both denominators of Eqs. (137) and (138) are quadratic in m of the form $am^2 + bm + c = 0$. For random frequency,

$$a = \gamma\lambda^3 - 4\omega^2\lambda^2\sigma^2; \quad b = 4\gamma\lambda\omega^2 + 3\gamma^2\lambda^2 - 4\gamma\lambda\omega^2\sigma^2;$$

$$c = 4\omega^2\gamma^2 + 2\lambda\gamma^3 - \omega^2\gamma^2\sigma^2 \qquad (141)$$

and for random damping,

$$a = \lambda^3; \quad b = 2\gamma\omega^2 + 3\gamma\lambda^2 - 2\gamma\lambda^2\sigma^2; \quad c = 2\gamma[\gamma\lambda(1 - \sigma^2) - 2\omega^2\sigma^2] \qquad (142)$$

The solutions of these equations are

$$m_{1,2} = \frac{-b \pm \sqrt{b^2 - 4ac}}{2a} \qquad (143)$$

For $b^2 - 4ac > 0$ and $c < 0$, both m_1 and m_2 are real with $m_2 < 0$. The system is unstable for $m_2 < m < m_1$ and stable for m outside this region. If $b^2 - 4ac > 0$ but $c > 0$, then m_1 is positive and m_2 is negative, and the system is unstable for $m < m_1$.

The mass m of an oscillator is real and positive. We are interested in the "energetic instability" of a system, which corresponds to a negative second moment $\langle x^2 \rangle$. The latter occurs when one of the ratios, say b_1/a_1, is negative, which means that both m_1 and m_2 are positive, and the other ratio b_2/a_2 is positive. Then, the denominator in Eq. (138) is positive for positive m and one of the factors in the numerator, $m - m_2$, is positive while the other factor, $m - m_1$ is negative for $m < m_1$. Another possibility of an instability is $b_1/a_1 > 0$ and $b_2/a_2 < 0$, which leads to instability for $m < m_2$. Therefore, in some cases (positive discriminants) an instability occurs when the masses are too small ($m < m_2$ or $m < m_1$).

For white multiplicative noise $\xi(t)$ with $\langle \xi(t_1)\xi(t_2) \rangle = D\delta(t_2 - t_1)$, the stationary second moments (137) and (138) are given by [28]

$$\langle x^2 \rangle_\omega = \frac{\alpha/\omega^2}{2\omega^2/m(\gamma/m - D\omega^2/m)} = \frac{\alpha}{2\omega^2(\gamma - D^2)};$$

$$\langle x^2 \rangle_\eta = \frac{\alpha/m^2}{\gamma\omega^2/m^2(1 - 2\gamma D/m)} = \frac{\alpha}{\gamma\omega^2(m - 2\gamma D)} \qquad (144)$$

As distinct from the usual analysis, we find the values of an oscillator mass for which a system becomes unstable. It turns out that for white multiplicative noise, the instability of the system with random damping occurs for an oscillator with a small mass, namely, $m < 2\gamma D$, but there are no restrictions on the system with random frequency.

The most striking phenomenon of these models with multiplicative noise is the existence of the transitions with no steady state ("energetic instability") for some values of the mass. It has been traditional in studies of harmonic oscillators subjected to multiplicative noise to scale out the mass by setting it equal to unity. This is of course possible, and one can write the critical noise amplitude (which is dimensionless) in terms of the dimensionless parameters $k/(m\lambda^2)$ and $\gamma/(m\lambda)$, where λ is the inverse correlation time of the noise.

Hence, even in the well-known cases of random frequency and random damping, the mass of the oscillator (or Brownian particle) is a very important parameter, which may lead to the instability of the system.

Systems with time-varying mass can be found in a variety of mechanical problems in astronomy, biology, economics, robotics and engineering (see, for example, the recent review of Cveticanin [81]). The equation of motion of an oscillator with a time-dependent mass $m(t)$ is

$$\frac{d}{dt}\left[m(t)\frac{dx}{dt}\right] + \gamma\frac{dx}{dt} + kx = \dot{f}(t) \tag{145}$$

Introducing the variable τ by means of

$$\tau = \tau(t) = \int_0^t \frac{dl}{m(l)} \tag{146}$$

changes the equation to the following form,

$$\frac{d^2x}{d\tau^2} + \gamma\frac{dx}{d\tau} + m[t(\tau)]x = m[t(\tau)]f[t(\tau)] \tag{147}$$

This procedure reduces the problem of an oscillator with a time-varying mass to the well-known problem of an oscillator with time-varying frequency. For a periodic function $m(t)$, Eq. (147) is the Hill equation.

2.6 Dichotomous random mass

We start with the calculation of the averaged first moment $\langle x \rangle$ for symmetric dichotomous noise (100). To perform the splitting of correlators, we use the well-known Shapiro-Loginov procedure [82] which yields, for exponentially correlated noise,

$$\frac{d}{dt}\langle \xi g \rangle = \left\langle \xi \frac{dg}{dt} \right\rangle - \lambda \langle \xi g \rangle \tag{148}$$

or

$$\left\langle \xi \frac{dg}{dt} \right\rangle = \left(\frac{d}{dt} + \lambda \right) \langle \xi g \rangle \tag{149}$$

and

$$\left\langle \xi \frac{d^n g}{dt^n} \right\rangle = \left(\frac{d}{dt} + \lambda \right)^n \langle \xi g \rangle \tag{150}$$

where g is an arbitrary function of ξ.

Since dichotomous noise is a linear function, one can write

$$\frac{1}{1 + \xi(t)} = \frac{1 - \xi(t)}{1 - \sigma^2} \tag{151}$$

which transforms Eq. (89) (with $m = 1$ and additive random force $\eta(t)$) into the following form

$$(1 - \sigma^2)\frac{d^2 x}{dt^2} + (1 - \xi(t)) \left(2\gamma \frac{d}{dt} + \omega^2 \right) x = \eta(t)(1 - \xi(t)) \tag{152}$$

Since the oscillator mass is positive, the condition $\xi^2 < 1$, and hence $\sigma^2 < 1$, will be satisfied in Eq. (152). For small oscillations of the mass, $\sigma^2 \ll 1$ and $\langle \eta \xi \rangle = 0$, Eq. (152) can be rewritten as

$$\frac{d^2 x}{dt^2} + \left(2\gamma \frac{d}{dt} + \omega^2 \right) x - \xi(t) \left(2\gamma \frac{d}{dt} + \omega^2 \right) x = \eta(t) \tag{153}$$

Averaging Eq. (152) and using Eq. (148) with $g = x$, leads to

$$(1 - \sigma^2)\frac{d^2 \langle x \rangle}{dt^2} + \left(2\gamma \frac{d}{dt} + \omega^2 \right) \langle x \rangle - \left(2\gamma \frac{d}{dt} + 2\gamma\lambda + \omega^2 \right) \langle \xi x \rangle = 0 \tag{154}$$

A new function $\langle \xi x \rangle$ enters Eq. (154). One can obtain a second equation for the two functions $\langle x \rangle$ and $\langle \xi x \rangle$ by multiplying Eq. (152) by $\xi(t)$ and averaging, using again Eqs. (148) and (149) with $g = x$ and $g = dx/dt$,

$$\left[(1 - \sigma^2) \left(\frac{d}{dt} + \lambda \right)^2 + 2\gamma \frac{d}{dt} + \lambda + \omega^2 \right] \langle \xi x \rangle$$

$$- \sigma^2 \left(2\gamma \frac{d}{dt} + \omega^2 \right) \langle x \rangle = 0 \tag{155}$$

The use of dichotomous noise offers a major advantage over other types of colored noise by terminating an infinite set of higher-order correlations, using the fact that $\xi(t)^2 = \sigma^2$. Eliminating $\langle \xi x \rangle$ from Eqs. (154) and (155), one obtains a cumbersome equation for the first moment $\langle x(t) \rangle$, which in the limiting case of white noise ($\lambda \to 0$, $\sigma^2 \to \infty$, so that $\sigma^2 \lambda = D$), reduces to

$$\frac{d^2 \langle x \rangle}{dt^2} + 2\gamma(1 - 2\gamma D) \frac{d \langle x \rangle}{dt} + \omega^2 (1 - 2\gamma D) \langle x \rangle = 0 \tag{156}$$

From $\sigma^2 < 1$, it follows that $2\gamma D < 1$, and the fluctuations of mass leads to small corrections to the frequency and the damping coefficient. The latter is connected to the restriction $\sigma^2 < 1$, which is absent for random frequency and damping coefficient.

Let us proceed to the analysis of the averaged second moment $\langle x^2 \rangle$, which has the following form [28]

$$\langle x^2 \rangle = \frac{4\alpha}{m\omega^2} \frac{[2\lambda(2\gamma + m\lambda)^2 + 8\omega^2(\gamma + \lambda m)] - m\sigma^2[\lambda(2\gamma + \lambda m) + \omega^2]}{8\gamma[\lambda(2\gamma + m\lambda)^2 + 4\omega^2(\gamma + \lambda m)] - m\lambda^2\sigma^2\omega^2} \tag{157}$$

The second moment $\langle x^2 \rangle$ becomes negative, implying an instability of the system, which occurs when the numerator or the denominator in Eq. (157) is negative, which takes place for noise strength larger than $(\sigma_0^2)_{m,1}$ or $(\sigma_0^2)_{m,2}$,

$$(\sigma_0^2)_{m,1} = \frac{2\lambda(2\gamma + m\lambda)^2 + 8\omega^2(\gamma + \lambda m)}{m[\lambda(2\gamma + \lambda m) + \omega^2]} \tag{158}$$

and

$$(\sigma_0^2)_{m,2} = \frac{8\gamma[\lambda(2\gamma + m\lambda)^2 + 4\omega^2(\gamma + \lambda m)]}{m\lambda^2\omega^2} \tag{159}$$

It is remarkable that if both inequalities $\sigma^2 > (\sigma_0^2)_{m,1}$ and $\sigma^2 > (\sigma_0^2)_{m,2}$ are satisfied, $\langle x^2 \rangle$ remains positive and the system is stable. Eq. (157) can be rewritten as

$$\langle x^2 \rangle = \frac{4\alpha[a_1 m^2 + b_1 m + c_1]}{m\omega^2[a_2 m^2 + b_2 m + c_2]} \approx \frac{(m - m_1)(m - m_2)}{(m - m_3)(m - m_4)} \tag{160}$$

where

$$a_1 = 2\lambda^3 - \lambda^2\sigma^2; \quad b_1 = \lambda(8\gamma\lambda + 2\omega^2) - \sigma^2(\omega^2 + 2\gamma\lambda);$$

$$c_1 = 4\gamma(\omega^2 + \lambda\gamma) \tag{161}$$

$$a_2 = 4\gamma\lambda^3; \quad b_2 = 32\lambda^2\gamma^2 + 32\lambda\gamma\omega^2 - \lambda^2\sigma^2\omega^2;$$

$$c_2 = 4\gamma(\omega^2 + 2\lambda\gamma^2) \tag{162}$$

The analysis of the "energetic instability" (negative second moment $\langle x^2 \rangle$) of a system with parameters (161)–(162) follows in perfect analogy to the analysis of the previous section. Since both c_1 and c_2 are positive, the determinants of the quadratic equations $b_{1,2}^2 - 4a_{1,2}c_{1,2}$ have to be positive for real masses. Moreover, it follows from Eqs. (160) that for positive ratio b/a, both values of $m_{1,2}$ (or $m_{3,4}$) are negative, whereas for negative b/a, one root is negative and the other root is positive. The instability occurs when one of the ratios, say b_1/a_1, is negative, which means that both m_1 and m_2 are positive, and the other ratio b_2/a_2 is positive. Then, the denominator in Eq. (160) is positive for positive m and one of the factors in the numerator, $m - m_2$, is positive, while the other factor, $m - m_1$ is negative for $m < m_1$. Another possibility of an instability is positive b_1/a_1 and negative b_2/a_2, which leads to an instability for $m < m_3$. Therefore, in some cases (positive discriminants) an instability occurs for masses that are very small ($m < m_3$ or $m < m_1$).

2.7 Stability of an oscillator with random mass

There are many publications concerning multiplicative sources of noise leading to random frequency and damping [28]. As distinguished from these analyses, we here consider the random mass case [27], which is described by the following dynamic equation (with $m = 1$) ,

$$[1 + \xi(t)]\frac{d^2 x}{dt^2} + \gamma\frac{dx}{dt} + \omega^2 x = \eta(t) \qquad (163)$$

with non-correlated additive $\eta(t)$ and multiplicative $\xi(t)$ noise with correlators

$$\langle \eta(t_1)\eta(t_2)\rangle = D\delta(|t_1 - t_2|); \quad \langle \xi(t_1)\xi(t_2)\rangle = \sigma^2\lambda\exp(-\lambda|t_1 - t_2|) \qquad (164)$$

The Langevin equation (163) for the random quantity $x(t)$ can be transformed to an equation for the averaged quantity $\langle x\rangle$ through the use of the Shapiro-Loginov transformations [82]. Multiplying Eq. (163) by ξ and $1 - \xi$ and averaging, one gets equations for $\langle x\rangle$ and for $\langle \xi x\rangle$. Eliminating $\langle \xi x\rangle$ from these two equations gives,

$$a_4\frac{d^4\langle x\rangle}{dt^4} + a_3\frac{d^3\langle x\rangle}{dt^3} + a_2\frac{d^2\langle x\rangle}{dt^2} + a_1\frac{d\langle x\rangle}{dt} + a_0\langle x\rangle = 0 \qquad (165)$$

where

$$a_4 = (1 - \sigma^2); \quad a_3 = 2[\gamma + \lambda(1 - \sigma^2)];$$

$$a_2 = [2\omega^2 + 3\lambda\gamma + \gamma^2 + \lambda^2(1 - \sigma^2)];$$

$$a_1 = (\gamma + \lambda)(2\omega^2 + \gamma\lambda); \quad a_0 = \omega^2(\omega^2 + \lambda^2 + \lambda\gamma) \qquad (166)$$

For the stationary states $(d/dt \ldots = 0)$, Eq. (165) gives $\langle x\rangle = 0$, as expected for the random quantity. However, the question arises of whether for all values of the parameters γ and ω^2 of the Brownian particle and σ and λ of the noise, the dynamic equation (165) will have the asymptotic $(t \to \infty)$ solution $\langle x\rangle = 0$. Perhaps this equation will not have finite solutions, meaning that the system is unable to reach the steady state. This is an object of our analysis to answer this question.

Let us seek the solution of Eq. (165) of the form $\langle x\rangle \simeq \exp(rt)$. This yields an algebraic equation of the form

$$a_4 r^4 + a_3 r^3 + a_2 r^2 + a_1 r + a_0 = 0 \qquad (167)$$

In the absence of multiplicative noise, $\xi(t) = 0$, substitution of $\langle x \rangle \simeq \exp(rt)$ in the averaged equation (163) results in

$$r^2 + \gamma r + \omega^2 = 0 \qquad (168)$$

which gives

$$r = -\frac{\gamma}{2} \pm \left(\frac{\gamma^2}{4} - \omega^2 \right)^{1/2} \qquad (169)$$

and $\langle x \rangle \simeq \exp(-|r|t)$, so that for $t \to \infty$, $\langle x \rangle \to 0$, as expected.

The situation is more complicated for the case of a random mass, which is described by Eq. (167). To examine the stability one has to calculate (167) for each set of parameters γ, ω^2, σ and λ. A comprehensive analysis of the quartic equation (167) exists in the literature. For nonzero values of all parameters and for not too large value of ω, for some sets of parameters, r becomes positive, which means that the oscillator with a random mass will not come to equilibrium. Referring the reader to the literature for the details [83], we explain here the main idea of the analysis.

The substitution $r = y - a_3/a_4$ transforms Eq. (167) to the simple (depressed) form

$$y^4 + py^2 + qy + r = 0, \qquad (170)$$

where

$$p = \frac{8a_4 - 3a_3^2}{8a_4^2}; \quad q = \frac{8a_1a_4 + a_3^3 - 4a_2a_3a_4}{8a_4^3};$$

$$r = \frac{16a_4a_3^2 - 64a_4^2a_3a_1 - 3a_3^4 + 256}{256a_4^4} \qquad (171)$$

Consider now equation

$$z^3 + pz^2 + \frac{p^2 - 4r}{4}z - \frac{q^2}{8} = 0 \qquad (172)$$

For $q \neq 0$ the left-hand side is negative at $z = 0$, positive at $z = \infty$, and always has a positive root z_0. The roots of the original quartic equation (167) are [40]

$$r_{1,...4} = \frac{1}{\sqrt{2}} \left\{ \left[\pm\sqrt{2z_0} - \sqrt{2z_0 - 4\left[\frac{p}{2} + z_0 \pm \frac{q}{2\sqrt{2z_0}} \right]} \right] \right\} \qquad (173)$$

Therefore, for $q \neq 0$, at least one positive root of Eq. (167) exists, which means the oscillator with a random mass becomes unstable. Therefore, the instability of an oscillator with a random mass is the rule rather than the exception. The limiting value $q = 0$ occurs only for very small noise strength, $\sigma^2 \ll 1$, when one can replace $1 - \sigma^2$ by unity.

It should be noted that a rigorous mathematical analysis of the stability of a stochastic differential equation is determined by the sign of the Lyapunov index. Such an analysis for an oscillator with a random mass has been performed [84].

We conclude that the random mass of an oscillator may lead to the divergence of the first moment, indicating that the system is unable to reach a steady state.

2.8　Stability conditions

Here we consider the more complicated problem of the stability of the solutions. For a deterministic equation, the stability of the fixed points is defined by the sign of β, found from the solution of the form $\exp(\beta t)$ of a linearized equation near the fixed points. The situation is quite different for a stochastic equation. The first moment $\langle x(t) \rangle$ and higher moments become unstable for some values of the parameters. However, the usual linear stability analysis, which leads to instability thresholds, turns out to be different for different moments. This makes them unsuitable for a stability analysis. A rigorous mathematical analysis of random dynamic systems shows [85] that, similar to the order–deterministic chaos transition in nonlinear deterministic equations, the stability of a stochastic differential equation is determined by the sign of Lyapunov exponents λ, which is defined as

$$\lambda = \frac{1}{2} \left\langle \frac{\partial \ln(x^2)}{\partial t} \right\rangle = \left\langle \frac{\partial x / \partial t}{x} \right\rangle \qquad (174)$$

Therefore, for stability analysis, one has to go from the quadratic Langevin-type equations to the associated Fokker-Planck equations with the appropriate coefficients, which describe the properties of statistical ensembles, and then to calculate the Lyapunov exponent λ.

One can see from Eq. (174) that it is convenient to replace the variable x in the Langevin equations by the variable $z = (dx/dt)/x$,

$$\frac{dz}{dt} = \frac{d^2x/dt^2}{x} - \frac{(dx/dt)^2}{x^2} \equiv \frac{d^2x/dt^2}{x} - z^2 \qquad (175)$$

The Lyapunov index λ now takes the following form [86]

$$\lambda = \int_{-\infty}^{\infty} zP_{st}(z)dz \qquad (176)$$

where $P_{st}(z)$ is the stationary solution of the Fokker-Planck equations corresponded to the Langevin equations expressing in the variable z.

For the foregoing reasons, white noise cannot be used for an oscillator with a random mass. We bring here the calculation for symmetric dichotomous noise $\xi = \pm\sigma$, described by Eq. (89) with $m = 1$. Replacing the variable x in Eq. (89) by the variable z, one obtains

$$\frac{dz}{d\tau} = A(z) + \xi B(z) \qquad (177)$$

where

$$A(z) = -z^2 - B(z); \quad B(z) = \frac{\gamma z + \omega^2}{1 + \sigma^2} \qquad (178)$$

According to [87], the stationary solution of the Fokker-Planck equation corresponding to the Langevin equation (177) has the following form

$$P_{st}(z) = N\frac{B}{\sigma^2 B^2 - A^2}$$

$$* \exp\left[-\frac{1}{2\tau}\int^z dx \left[\frac{1}{A(x) - \sigma B(x)} + \frac{1}{A(x) + \sigma B(x)}\right]\right]$$

$$(179)$$

where N is the normalization constant.

The zeroes of functions $F_\pm(x) = \pm\sigma B(x) - A(x)$ determine the boundary of support of $P_{st}(z)$, which diverges or vanishes at the boundaries. The latter means that a system will approach the state z located in intervals (z_2, z_1) or (z_4, z_3), depending on its initial position. Another important characteristic of $P_{st}(z)$ is the location of its extrema, which define the macroscopic steady states. The steady states x_m of (179) obey the following equation [87]

$$A x_m - \sigma^2 \tau B(x_m) \frac{d}{dx} A(x_m) + 2\tau A(x_m) \frac{d}{dx} A(x_m)$$

$$- \tau \frac{A(x_m)^2 B(x_m)}{dB(x_m)/dx} = 0 \tag{180}$$

where

$$A = \alpha x^2 + \beta x + \kappa; \quad B = \beta x + \kappa \tag{181}$$

with

$$\alpha = -1; \quad \beta = -\frac{\gamma}{R}(1 + \sigma^2); \quad \kappa = -\frac{\omega^2}{R}(1 + \sigma^2) \tag{182}$$

The first term in (180) defines the deterministic steady states, whereas the last two terms define the corrections coming from the invert correlation time λ.

Inserting (181) into (179), one obtains

$$P_{st}(z) = N(z - x_1)^{-1-[2\lambda\alpha(x_1-x_2)]^{-1}} (z - x_2)^{-1+[2\lambda\alpha(x_1-x_2)]^{-1}}$$

$$* (z - x_3)^{-1+[2\lambda(x_3-x_4)]^{-1}} (z - x_4)^{-1-[2\lambda(x_3-x_4)]^{-1}} \tag{183}$$

where

$$x_{1,2} = -\frac{(1+\sigma)\beta}{2\alpha} \pm \sqrt{\left(\frac{(1+\sigma)\beta}{2\alpha}\right)^2 - \frac{(1+\sigma)\kappa}{\alpha}}$$

$$= -\gamma Q_+ \pm \sqrt{\gamma^2 Q_+^2 - \omega^2 Q_+} \tag{184}$$

and

$$x_{3,4} = -\frac{(1-\sigma)\beta}{2\alpha} \pm \sqrt{\left(\frac{(1-\sigma)\beta}{2\alpha}\right)^2 + \frac{(1-\sigma)\kappa}{\alpha}}$$

$$= -\gamma Q_- \pm \sqrt{\gamma^2 Q_-^2 - \omega^2 Q_-} \tag{185}$$

with

$$Q_\pm = \frac{(1+\sigma^2)(1 \pm \sigma)}{R} \tag{186}$$

Equation (183) defines the boundary of stability of the fixed point $x = 0$ for different values of parameters $\gamma Q_\pm$ and $\omega^2 Q_\pm$, which depend on the characteristics ω^2, γ of the oscillator and σ, Δ and λ of the noise.

2.9 Basic equations

For generality we consider trichotomous noise. For the stationary states, the probabilities P of values $\pm a$ and 0 are

$$P(-a) = P(a) = q; \quad P(0) = 1 - 2q \tag{187}$$

The limiting case of dichotomous noise corresponds to $q = 1/2$. The supplementary conditions to the Orenstein-Uhlenbeck correlations are

$$\xi^3 = a^2\xi; \quad \xi^2 = 2qa^2 \tag{188}$$

Equation (40) with $m = 1$ in the external field $A\sin(\Omega t)$ can be rewritten as two first-order differential equations

$$\frac{dx}{dt} = y; \quad \frac{dy}{dt} = -\xi\frac{dy}{dt} - \gamma y - \omega^2 x + A\sin(\Omega t) + \eta(t) \tag{189}$$

which take the following form after averaging

$$\frac{d\langle x \rangle}{dt} = \langle y \rangle;$$

$$\frac{d\langle y \rangle}{dt} = -\left(\frac{d}{dt} + \lambda\right)\langle \xi y \rangle - \gamma\langle y \rangle - \omega^2\langle x \rangle + A\sin(\Omega t) \tag{190}$$

where we have used the Shapiro-Loginov formula (149) for splitting the correlation.

Additional relations between averaged values can be obtained by multiplying the first of Eq. (189) by $2x$ and the second by $2y$,

$$\frac{d}{dt}x^2 = 2xy; \quad \frac{d}{dt}y^2 + \xi\frac{dy^2}{dt} + 2\gamma y^2 + 2\omega^2 xy = 2yA\sin(\Omega t) \tag{191}$$

Averaging Eqs. (191) yields

$$\frac{d}{dt}\langle x^2 \rangle = 2\langle xy \rangle;$$

$$\left(\frac{d}{dt} + 2\gamma\right)\langle y^2 \rangle + \left(\frac{d}{dt} + \lambda\right)\langle \xi y^2 \rangle + 2\omega^2\langle xy \rangle = 2\langle y \rangle A\sin(\Omega t) \tag{192}$$

Analogously, multiplying Eqs. (189) by y and x, respectively, and summing leads to

$$\frac{d}{dt}xy = y^2 - \xi\left[\frac{d}{dt}(xy) - y^2\right] - \gamma xy - \omega^2 x^2 + 2xA\sin(\Omega t)$$

$$= y^2 - \left(\frac{d}{dt} + \lambda\right)\xi xy + \xi y^2 - \gamma xy - \omega^2 x^2 + 2xA\sin(\Omega t) \tag{193}$$

which yields, after averaging,

$$\frac{d}{dt}\langle xy\rangle = \langle y^2\rangle - \left(\frac{d}{dt} + \lambda\right)\langle\xi xy\rangle$$

$$+ \langle\xi y^2\rangle - \gamma\langle xy\rangle - \omega^2\langle x^2\rangle + 2\langle x\rangle A\sin(\Omega t) \tag{194}$$

Additional equations for the correlators can be obtained by multiplying Eqs. (189) and (193) by $2\xi x, 2\xi y$ and ξ, respectively, and averaging,

$$\left(\frac{d}{dt} + \lambda\right)\langle\xi x^2\rangle = 2\langle\xi xy\rangle \tag{195}$$

$$\left(\frac{d}{dt} + \lambda + 2\gamma\right)\langle\xi y^2\rangle + 2\omega^2\langle\xi xy\rangle = 2\langle\xi y\rangle A\sin(\Omega t) \tag{196}$$

$$\left(\frac{d}{dt} + \lambda + \gamma\right)\langle\xi xy\rangle = \langle\xi y^2\rangle - \omega^2\langle\xi x^2\rangle + \omega^2\sigma^2\langle x^2\rangle$$

$$+ 2\langle\xi y\rangle A\sin(\Omega t) \tag{197}$$

which, in the long-time limit, $t \to \infty$, are reduced to

$$\lambda\langle\xi x^2\rangle = 2\langle\xi xy\rangle; \quad (\lambda + 2\gamma)\langle\xi y^2\rangle + 2\omega^2\langle\xi xy\rangle = 2\langle\xi y\rangle A\sin(\Omega t);$$

$$(\lambda + \gamma)\langle\xi xy\rangle = \langle\xi y^2\rangle - \omega^2\langle\xi x^2\rangle + \omega^2\sigma^2\langle x^2\rangle + 2\langle\xi y\rangle A\sin(\Omega t) \tag{198}$$

In the limit $t \to \infty$, Eqs. (194), (195), (196) and (197) take the following form

$$\lambda\langle\xi y\rangle + \omega^2\langle x\rangle = A\sin(\Omega t) \tag{199}$$

$$2\gamma\langle y^2\rangle + \lambda\langle\xi y^2\rangle = 0 \tag{200}$$

$$\langle y^2\rangle - \lambda\langle\xi xy\rangle + \langle\xi y^2\rangle - \gamma\langle xy\rangle - \omega^2\langle x^2\rangle + 2\langle x\rangle A\sin(\Omega t) = 0 \tag{201}$$

Multiplying Eqs. (189) by $\xi(t)$ and averaging results in

$$\left(\frac{d}{dt} + \lambda\right)\langle\xi x\rangle = \langle\xi y\rangle;$$

$$\left(\frac{d}{dt} + \lambda\right)\langle\xi y\rangle = -\left\langle\xi^2\left[-\xi\frac{dy}{dt} - \gamma y - \omega^2 x + A\sin(\Omega t)\right]\right\rangle$$

$$-\gamma\langle\xi y\rangle - \omega^2\langle\xi x\rangle$$

$$= a^2\left(\frac{d}{dt} + \lambda\right)\langle\xi y\rangle + \gamma\langle\xi^2 y\rangle + \omega^2\langle\xi^2 x\rangle$$

$$- 2qAa^2\sin(\Omega t) - \gamma\langle\xi y\rangle - \omega^2\langle\xi x\rangle; \qquad (202)$$

Finally, for the stationary states, Eqs. (202) take the form

$$\langle\xi^2 y\rangle - \lambda\langle\xi^2 x\rangle + 2q\lambda a^2\langle x\rangle = 0 \qquad (203)$$

and

$$-a^2\lambda\langle\xi y\rangle - (\gamma + \lambda)\langle\xi^2 y\rangle - \omega^2\langle\xi^2 x\rangle + 2q\lambda a^2\langle y\rangle + 2qAa^2\sin(\Omega t) = 0 \qquad (204)$$

By this means we obtain ten equations (198), (199), (115), (201), (200), (204) for the ten correlators $\langle x\rangle$, $\langle x^2\rangle$, $\langle y^2\rangle$, $\langle\xi x\rangle$, $\langle\xi y\rangle$, $\langle\xi xy\rangle$, $\langle\xi x^2\rangle$, $\langle\xi y^2\rangle$, $\langle\xi^2 x\rangle$ and $\langle\xi^2 y\rangle$.

2.10 First moment

In order to find $\langle x\rangle$ from the equations obtained in the previous section, let us first eliminate $\langle\xi y\rangle$ from these equations, which gives

$$\lambda^2\langle\xi x\rangle + \omega^2\langle x\rangle = A\sin(\Omega t) \qquad (205)$$

$$\lambda[\lambda(1 - a^2) + \gamma] + \omega^2\langle\xi x\rangle = \gamma\langle\xi^2 y\rangle + \omega^2\langle\xi^2 x\rangle - 2qAa^2\sin(\Omega t) \qquad (206)$$

$$2\gamma\langle y^2\rangle + \lambda\langle\xi^2 y\rangle = A\sin(\Omega t) \qquad (207)$$

$$\lambda\langle\xi x^2\rangle = 2\langle\xi xy\rangle;$$

$$(\lambda + 2\gamma)\langle\xi y^2\rangle + 2\omega^2\langle\xi xy\rangle = 2\lambda\langle\xi x\rangle A\sin(\Omega t) \qquad (208)$$

Inserting $\langle \xi^2 y \rangle$ from (206) into (207) yields

$$\frac{\lambda\{\lambda[\lambda(1-a^2)+\gamma]+\omega^2\}}{\gamma}\langle \xi x \rangle - \frac{\omega^2\lambda}{\gamma}\langle \xi^2 x \rangle + 2\gamma\langle y^2 \rangle$$

$$+\frac{2q\lambda Aa^2}{\gamma}\sin(\Omega t) = A\sin(\Omega t) \tag{209}$$

$$-\left\{\frac{(2\gamma+\lambda)\left\{\lambda\left[\lambda(1-a^2)+\gamma\right]+\omega^2\right\}}{\gamma}\right\}\langle \xi x \rangle$$

$$-\frac{\omega^2(2\gamma+\lambda)}{\gamma}\langle \xi^2 x \rangle - \frac{2qa^2(2\gamma+\lambda)}{\gamma}A\sin(\Omega t) + \omega^2\lambda\langle \xi x^2 \rangle$$

$$= 2\lambda\langle \xi x \rangle A\sin(\Omega t) \tag{210}$$

Substituting $\langle \xi x \rangle$ from (205) into (209) and (210) leads to

$$\frac{\omega^2\{\lambda[\lambda(1-a^2)+\gamma]+\omega^2\}}{\gamma\lambda}\langle x \rangle + \frac{\{\lambda[\lambda(1-a^2)+\gamma]+\omega^2\}}{\gamma\lambda}A\sin(\Omega t)$$

$$-\frac{\omega^2\lambda}{\gamma}\langle \xi^2 x \rangle + 2\gamma\langle y^2 \rangle + \frac{2q\lambda a^2}{\gamma}A\sin(\Omega t) = A\sin(\Omega t) \tag{211}$$

$$\frac{\omega^2(2\gamma+\lambda)\{\lambda[\lambda(1-a^2)+\gamma]+\omega^2\}}{\gamma\lambda^2}\langle x \rangle$$

$$-\frac{(2\gamma+\lambda)\{\lambda[\lambda(1-a^2)+\gamma]+\omega^2\}}{\gamma\lambda^2}A\sin(\Omega t)$$

$$-\frac{\omega^2(2\gamma+\lambda)}{\gamma}\langle \xi^2 x \rangle - \frac{2qa^2(2\gamma+\lambda)}{\gamma}A\sin(\Omega t)$$

$$+\omega^2\lambda\langle \xi x^2 \rangle = 2\lambda\langle \xi x \rangle A\sin(\Omega t) \tag{212}$$

Finally, eliminating $\langle \xi^2 x \rangle$ from (211) and (212) leads to

$$\langle x \rangle = \frac{\lambda^3 D_1}{\gamma\omega^2\{B\omega^2\lambda - \gamma\lambda(\omega^2+\lambda\gamma)[B(\lambda+\gamma)-a^2\lambda^2]-2q\lambda^4 a^2\}}$$

$$+\left\{\frac{2q\lambda a^2}{\omega^2} + \frac{B}{\lambda^2} - \frac{(\omega^2+\lambda\gamma)\left[B(\gamma+\lambda)-a^2\lambda^2\right]}{\omega^2\lambda^3\gamma}\right\}A\sin(\Omega t)$$

$$\tag{213}$$

where

$$B = \frac{\lambda\{[\lambda(1 - a^2) - \gamma] + \omega^2\}}{\gamma}$$

The first term in Eq. (213) describes the common action on an oscillator of the additive and multiplicative forces while the second term is related to the oscillator response to an external periodic force. As can be seen from (213), the first moment $\langle x \rangle$ shows non-monotonic dependence on the parameters of noise and periodic force. Moreover, for some values of these parameters, an oscillator becomes unstable.

2.11 White noise

Equation (89) with $m = 1$ can be rewritten in the following form

$$\frac{d^2x}{dt^2} + \gamma\frac{dx}{dt} + \omega^2 x = -\xi\frac{d^2x}{dt^2} \tag{214}$$

Based on linear response theory, the output $x(t)$ of the system for the input $-\xi d^2x/dt^2$ is

$$x(t) = \frac{1}{\omega_1}\int_0^t \exp\left[-\frac{\gamma}{2}(t - u)\right]\sin\left[\omega_1(t - u)\right]\left(-\xi\frac{d^2x(u)}{dt^2}\right) du \tag{215}$$

where $\omega_1 = \sqrt{\omega^2 - \gamma^2/4}$. Finding d^2x/dt^2 from Eq. (215), inserting it into Eq. (214) and using the well-known formula for splitting the correlations,

$$\left\langle \xi(t)\xi(t_1)\frac{d^2x}{dt}(t_1) \right\rangle = \langle \xi(t)\xi(t_1) \rangle \left\langle \frac{d^2x}{dt}(t_1) \right\rangle \tag{216}$$

one obtains for white noise,

$$(1 - \gamma D)\frac{d^2}{dt^2}\langle x \rangle + \gamma\frac{d}{dt}\langle x \rangle + \omega^2\langle x \rangle = 0 \tag{217}$$

which implies a renormalization of oscillator mass.

2.12 Symmetric dichotomous noise

Averaging Eq. (152) over an ensemble of random functions $\xi(t)$ and using Eq. (148) with $g = x$ leads to

$$(1 - \sigma^2)\frac{d^2\langle x\rangle}{dt^2} + \left(\gamma\frac{d}{dt} + \omega^2\right)\langle x\rangle - \left(\gamma\frac{d}{dt} + \gamma\lambda + \omega^2\right)\langle\xi x\rangle = -R$$

$$(218)$$

where we assume white-noise correlations of noise $\xi(t)$ and $\eta(t)$, that is, $\langle\xi(t)\eta(t_1)\rangle = R\delta(t - t_1)$.

The new function $\langle\xi x\rangle$ enters Eq. (218). One can obtain a second equation for the two functions $\langle x\rangle$ and $\langle\xi x\rangle$ by substituting Eq. (151) into (89) (with $m = 1$) and multiplying the obtained equation by $\xi(t)$.

Then, one gets after averaging and using Eq. (148) with $g = x$ and $g = dx/dt$,

$$\left[(1 - \sigma^2)\left(\frac{d}{dt} + \lambda\right)^2 + \gamma\frac{d}{dt} + \gamma\lambda + \omega^2\right]\langle\xi x\rangle$$

$$-\sigma^2\left(\gamma\frac{d}{dt} + \omega^2\right)\langle x\rangle = 0 \qquad (219)$$

The use of dichotomous noise offers a major advantage over other types of colored noise by terminating an infinite set of higher-order correlations, using the fact that $\langle\xi(t)^2\rangle = \sigma^2$. Eliminating $\langle\xi x\rangle$ from Eqs. (218) and (219), one obtains the following cumbersome equation for the first moment $\langle x(t)\rangle$

$$(1 - \sigma^2)\frac{d^4\langle x\rangle}{dt^4} + [2\gamma + 2\lambda(1 - \sigma^2)]\frac{d^3\langle x\rangle}{dt^3}$$

$$+ [2\omega^2 + \gamma^2 + 3\gamma\lambda + \lambda^2(1 - \sigma^2)]\frac{d^2\langle x\rangle}{dt^2}$$

$$+ (\lambda + \gamma)(\gamma\lambda + 2\omega^2)\frac{d\langle x\rangle}{dt} + \omega^2[\omega^2 + \lambda(\lambda + \gamma)]\langle x\rangle$$

$$= -R(\lambda^2(1 - \sigma^2) + \gamma\lambda + \omega^2) \qquad (220)$$

Equations (218) and (219) can also be solved by the Laplace transform,

$$\langle X(p) \rangle = \int_0^\infty \langle x(t) \rangle \exp(-pt)dt, \tag{221}$$

with initial conditions $x(t = 0) = X_0$, $dx/dt(t = 0) = (dX/dt)_0$, $\langle \xi x \rangle$ $(t = 0) = \langle \xi y \rangle$ $(t = 0) = 0$. For $\lambda = 0$, one obtains,

$$\langle X(p) \rangle = \frac{[M - \sigma^2 p + \lambda^2][pX_0 + (dX/dt)_0] + 2\gamma M X_0}{M(p^2 + 2p\gamma + \omega^2) - \sigma^2 p^2(p + \lambda)^2} \tag{222}$$

where $M = (p + \lambda)^2 + 2\gamma(p + \lambda) + \omega^2$.

Another way to analyze the differential equation (214) is to transform it into an integro-differential equation [88]. For this purpose, we rewrite the equation in the following form

$$L\{x\} = -\xi\frac{d^2 x}{dt^2}; \quad L\{x\} \equiv \frac{d^2}{dt^2} + 2\gamma\frac{d}{dt} + \omega^2 \tag{223}$$

Applying the operator L^{-1} to the first of Eqs. (223), one obtains

$$x(t) = -L^{-1}\left\{\xi\frac{d^2 x}{dt^2}\right\} \tag{224}$$

Using the fact that $L\{L^{-1}\{f\}\} = f$, one can easily check that the integral operator L^{-1} is the inverse of the differential operator L, defined in (223), where L^{-1} has the following form

$$L^{-1}\{\psi(t)\} = \frac{1}{\omega_1}\int_0^t dt_1 \exp[-\gamma(t - t_1)]\sin[\omega_1(t - t_1)]\psi(t_1);$$

$$\omega_1 = \sqrt{\omega^2 - \gamma^2/4} \tag{225}$$

Using (224) and (225) yields

$$x(t) = -\frac{1}{\omega_1}\int_0^t dt_1 \exp[-\gamma(t - t_1)]\sin[\omega_1(t - t_1)]\xi(t_1)\frac{d^2 x}{dt^2}(t_1) \tag{226}$$

and

$$\frac{d^2x}{dt^2} = \xi(t)\frac{d^2x}{dt^2} - \frac{1}{\omega_1}\int_0^t dt_1\{(\omega^2 - 2\gamma^2)\sin[\omega_1(t-t_1)]$$

$$+ 2\omega_1\gamma\cos[\omega_1(t-t_1)]\}\exp[-\gamma(t-t_1)]\xi(t_1)\frac{d^2x}{dt^2}(t_1)$$

$$(227)$$

Multiplying Eq. (227) by $\xi(t)$, substituting the resulting expression into Eq. (223), and averaging over noise, one obtains

$$\frac{d^2}{dt^2}\langle x\rangle + 2\gamma\frac{d}{dt}\langle x\rangle + \omega^2\langle x\rangle$$

$$= \langle\xi^2(t)(d^2x/dt^2)\rangle - \frac{1}{\omega_1}\int_0^t dt_1\{(\omega^2 - 2\gamma^2)\sin[\omega_1(t-t_1)]$$

$$+ 2\omega_1\gamma\cos[\omega_1(t-t_1)]\}\exp[-\gamma(t-t_1)]\langle\xi(t)\xi(t_1)\rangle\frac{d^2}{dt^2}\langle x\rangle(t_1)$$

$$(228)$$

For the analysis of Eq. (228), we use Eq. (216),

$$\frac{d^2}{dt^2}\langle x\rangle + 2\gamma\frac{d}{dt}\langle x\rangle + \omega^2\langle x\rangle$$

$$= \sigma^2\frac{d^2}{dt^2}\langle x\rangle - \frac{\sigma^2}{\omega_1}\int_0^t dt_1\{(\omega^2 - 2\gamma^2)\sin[\omega_1(t-t_1)]$$

$$+ 2\omega_1\gamma\cos[\omega_1(t-t_1)]\}\exp[-(\lambda+\gamma)(t-t_1)](d^2\langle x\rangle/dt^2)(t_1)$$

$$(229)$$

The integrand in Eq. (229) has the form $f(t_1)g(t-t_1)$. Using the convolution theorem for the Laplace transform [89],

$$L\left\{\int_0^t f(t_1)g(t-t_1)dt_1\right\} = L\{f\}L\{g\}, \qquad (230)$$

one gets Eq. (165), obtained earlier by the other method.

2.13 Asymmetric dichotomous noise

Starting from Eq. (95), using Eq. (150) with $n = 2$, and averaging over noise yields

$$\left[(1+\sigma^2)\frac{d^2}{dt^2} + \gamma\frac{d}{dt} + \omega^2\right]\langle x\rangle + \Delta\left(\frac{d}{dt} + \lambda\right)^2\langle\xi x\rangle = 0 \qquad (231)$$

A second equation for the two functions $\langle x\rangle$ and $\langle\xi x\rangle$ can be obtained from Eq. (101) by using Eq. (150) (with $n = 2$ and $n = 1$),

$$\left[R\frac{d^2}{dt^2}\langle x\rangle + (1+\sigma^2)\left(\gamma\frac{d}{dt} + \omega^2\right)\right]\langle x\rangle$$

$$-\left[\Delta^3\left(\frac{d}{dt} + \lambda\right)^2 + \Delta\gamma\left(\frac{d}{dt} + \lambda\right) + \Delta\omega^2\right]\langle\xi x\rangle = 0$$

$$(232)$$

where

$$\langle\xi(r_1)\eta(r_2)\rangle = R\delta(r_1 - r_2) \qquad (233)$$

Excluding the correlator $\langle\xi x\rangle$ from Eqs. (231) and (232), one obtains the following fourth-order differential equation for $\langle x\rangle$,

$$[(1+\sigma^2)^2 + \Delta^2]\frac{d^4}{dt^4}\langle x\rangle + [\Delta^2(\gamma + 2\lambda) + 2\lambda(1+\sigma^2)$$

$$+ [2\gamma(1+\sigma^2)]\frac{d^3}{dt^3}\langle x\rangle + [\Delta^2(\lambda + \gamma)$$

$$+ (1+\sigma^2)(2\omega^2 + \lambda + 3\gamma)$$

$$+ \omega^2\Delta^2 + \gamma^2]\frac{d^2}{dt^2}\langle x\rangle + [[\lambda(1+\sigma^2+\Delta^2) + \gamma]$$

$$\times (2\omega^2 + \lambda\gamma)\frac{d}{dt}\langle x\rangle$$

$$+ \omega^2[\omega^2 + \lambda\gamma + \lambda^2(1+\sigma^2+\Delta^2)]\langle x\rangle = 0 \qquad (234)$$

2.14 Second moment

For the calculation of the second moment $\langle x^2 \rangle$, we multiply the first of equations (189) by $2x$ and the second by $2y$

$$\frac{d}{dt}x^2 = 2xy;$$

$$\frac{d}{dt}y^2 + \xi\frac{dy^2}{dt} + 2\gamma y^2 + 2\omega^2 xy = 2y\eta(t) + 2Ay\sin(\Omega t) \tag{235}$$

Using the well-known formula for splitting the correlations, which is exact for the Ornstein-Uhlenbeck noise, one obtains

$$\langle \xi(t)\eta(t)y \rangle = \langle \xi(t)\eta(t) \rangle\langle y \rangle \tag{236}$$

Averaging Eqs. (235) yields

$$\frac{d}{dt}\langle x^2 \rangle = 2\langle xy \rangle;$$

$$\left(\frac{d}{dt} + 2\gamma\right)\langle y^2 \rangle + \left(\frac{d}{dt} + \frac{1}{\tau}\right)\langle \xi y^2 \rangle + 2\omega^2\langle xy \rangle$$

$$= 2D_1 + 2A\langle y \rangle\sin(\Omega t) \tag{237}$$

Analogously, multiplying Eqs. (189) by y and x, respectively, and summing leads to

$$\frac{d}{dt}xy = y^2 - \xi x\left(-\xi\frac{dy}{dt} - \gamma y - \omega^2 x + \eta(t)\right)$$

$$- \gamma xy - \omega^2 x^2 + \eta x + 2A\langle x \rangle\sin(\Omega t)$$

$$= y^2 + \sigma^2\left[\frac{d}{dt}(xy) - y^2\right] + \gamma\xi xy + \omega^2\xi x^2 - \xi\eta x - \gamma xy$$

$$- \omega^2 x^2 + x\eta + 2A\langle x \rangle\sin(\Omega t) \tag{238}$$

After averaging and using $\langle x\xi\eta \rangle = 0$, one obtains for $\sigma^2 < 1$,

$$\frac{d}{dt}\langle xy \rangle = \langle y^2 \rangle + \gamma\langle \xi xy \rangle + \omega^2\langle \xi x^2 \rangle - \gamma\langle xy \rangle - \omega^2\langle x^2 \rangle + 2A\langle x \rangle\sin(\Omega t) \tag{239}$$

Equations (237) and (239) contain new correlators $\langle \xi x^2 \rangle$, $\langle \xi y^2 \rangle$ and $\langle \xi xy \rangle$. One can calculate these correlators by multiplying Eqs.

(189) and (238) by $2\xi x, 2\xi y$ and ξ, respectively, and averaging,

$$\left(\frac{d}{dt} + \frac{1}{\tau}\right)\langle \xi x^2 \rangle = 2\langle \xi xy \rangle \tag{240}$$

$$\left(\frac{d}{dt} + \frac{1}{\tau} + 2\gamma\right)\langle \xi y^2 \rangle + 2\omega^2 \langle \xi xy \rangle = 2A\langle \kappa y \rangle \sin(\Omega t) \tag{241}$$

$$\left(\frac{d}{dt} + \frac{1}{\tau} + \gamma\right)\langle \xi xy \rangle = \langle \xi y^2 \rangle - \omega^2 \langle \xi x^2 \rangle + \omega^2 \sigma^2 \langle x^2 \rangle$$

$$+ \lambda \langle x \rangle + 2A\langle \xi x \rangle \sin(\Omega t) \tag{242}$$

By this means we obtain six equations, (237) and (239)–(242), for the six variables $\langle x^2 \rangle$, $\langle y^2 \rangle$, $\langle xy \rangle$, $\langle \xi x^2 \rangle$, $\langle \xi y^2 \rangle$, and $\langle \xi xy \rangle$. We will not write here the cumbersome dynamic equations for the second moments, which can be easily obtained from this system of differential equations, but shall restrict our attention to the case of small fluctuations of mass ($\sigma^2 < 1$) in the absence of an external periodic force, $A = 0$ and in the limiting case of the long-time limit, $t \to \infty$,

$$\langle x^2 \rangle = \frac{D_1}{\gamma \omega^2} \frac{1}{1 - \sigma^2 \alpha/\gamma} \tag{243}$$

where

$$\alpha = \frac{\omega^2 + \gamma(1 + 2\gamma\tau)(\gamma + 2\omega^2\tau)}{2\omega^2\tau + (2\omega^2\tau + \gamma + 1/\tau)(1 + 2\gamma\tau)}; \quad \eta(t_1, t_2) = D_1\delta(t_1 - t_2). \tag{244}$$

For white noise, Eq. (243) reduces to

$$\langle x^2 \rangle = \frac{D_1}{\gamma \omega^2} \tag{245}$$

This result coincides with the well-known result for "free" Brownian motion. For "free" Brownian particle, $\omega^2 \to 0$, one obtains $\langle x^2 \rangle \to \infty$, as required for Brownian motion. The independence of the stationary results on the mass fluctuation is due to the fact that the multiplicative random force appears in Eq. (89) in front of the higher derivative. It is remarkable that these results are significantly different from the stationary second moments for the white-noise random frequency

$\langle x^2 \rangle_\omega$ and random damping $\langle x^2 \rangle_\gamma$,

$$\langle x^2 \rangle_\omega = \frac{D_1}{2\omega^2(\gamma - D\omega^2)}; \quad \langle x^2 \rangle_\gamma = \frac{D_1}{\gamma\omega^2(1 - 2\gamma D)} \tag{246}$$

which exhibit an "energetic" instability [90]. It turns out that for symmetric dichotomous noise, the stationary second moment $\langle x^2 \rangle$ for the mass fluctuations, in contrast to its white noise form (245), may lead to instability, $(\langle x^2 \rangle < 0)$ for $\sigma^2 \alpha > \gamma$.

The correlation function can be found by multiplying equations (189) by $x(t_1)$ and averaging the resulting equations,

$$\frac{d}{dt}\langle x(t_1)x(t) \rangle = \langle x(t_1)y(t) \rangle$$

$$\frac{d}{dt}\langle x(t_1)y(t) \rangle = \frac{d}{dx}\langle \xi(t)x(t_1)y(t) \rangle - \gamma\langle x(t_1)y(t) \rangle - \omega^2\langle x(t_1)x(t) \rangle$$

$$\tag{247}$$

The new correlator $\langle \xi(t)\langle x(t_1)y(t) \rangle \rangle$ can be found by using Eq. (216),

$$\left(\frac{d}{dt} + \lambda \right) \langle \xi(t)x(t_1)x(t) \rangle = \langle \xi(t)x(t_1)y(t) \rangle;$$

$$\left(\frac{d}{dt} + \lambda \right)^2 \langle \xi(t)x(t_1)x(t) \rangle + \sigma^2\frac{d}{dt}\langle x(t_1)y(t) \rangle + \gamma\langle \xi(t)x(t_1)y(t) \rangle$$

$$+ \omega^2\langle \xi(t)x(t_1)x(t) \rangle = 0 \tag{248}$$

From Eqs. (247)–(248), one can find the fourth-order differential equation for the correlation function $\langle x(t_1)x(t) \rangle$, which, due to the linearity of this equation, coincides with equation (188) for the first moment.

For dichotomous noise, the correlation function shows a non-monotonic dependence on both the noise strength σ^2 and the inverse correlation time λ^{-1}.

2.15 Instability of the second moment

We here continue the analysis of the second moment considered in the previous section. For trichotomous multiple noise $\xi(t)$, which consists of jumps between three values $\pm a$ and zero, the stationary state correlation function has the following form,

$$\langle \xi(t)\xi(t+\tau)\rangle = 2qa^2 \exp(-\nu\tau), \tag{249}$$

Averaging the noise equations obtained in the previous section leads to [91],

$$\frac{d\langle x\rangle}{dt} = \langle y\rangle; \quad \frac{d\langle y\rangle}{dt} = -\left(\frac{d}{dt}+\nu\right)\langle \xi y\rangle - 2\gamma\langle y\rangle - \omega^2\langle x\rangle;$$

$$\frac{d\langle \xi x\rangle}{dt} = -\nu\langle \xi x\rangle + \langle \xi y\rangle;$$

$$\frac{d\langle \xi y\rangle}{dt} = -(\nu+2\gamma)\langle \xi y\rangle - \left(\frac{d}{dt}+\nu\right)\langle \xi^2 y\rangle - \omega^2\langle \xi x\rangle + 2q\nu a^2\langle y\rangle = 0;$$

$$\frac{d\langle \xi^2 x\rangle}{dt} = -\nu\langle \xi^2 x\rangle + \langle \xi^2 y\rangle + 2q\nu a^2\langle x\rangle = 0;$$

$$\frac{d\langle \xi^2 y\rangle}{dt} = -(\nu+2\gamma)\langle \xi^2 y\rangle - a^2\frac{d\langle \xi y\rangle}{dt} - \nu a^2\langle \xi y\rangle - \omega^2\langle \xi^2 x\rangle + 2q\nu a^2\langle y\rangle \tag{250}$$

The second moment $\langle x^2\rangle$ can be found from the system of six equations (250) for six variables. For stationary states, this gives

$$\langle x^2\rangle = \frac{2D\{[1-a^2(1-2q)]g_1 + qa^2\nu g_2\}}{(1-a^2)[4\gamma g_1 - qa^2\nu^2 g_3]} \tag{251}$$

where

$$g_1 = \{|4\omega^2 + \nu(\nu+4\gamma)|^2 - a^2\nu^4(1-2q)\}[(\nu+2\gamma)^2 - \nu^2 a^2(1-2q)]$$
$$+ qa^2\nu^3\{(\nu+2\gamma)[4\omega^2 + \nu(\nu+4\gamma)] + a^2\nu^3(1-2q)\};$$

$$g_2 = [\nu+2\gamma + a^2\nu(1-2q)]\{4\omega^2 + \nu(\nu+4\gamma)^2] - a^2\nu^4(1-2q)\};$$

$$g_3 = (\nu+2\gamma)\{[4\omega^2 + \nu(\nu+4\gamma)]^2 - a^2\nu^4(1-2q)\} + 2a^2\nu^4(1-2q)$$
$$- \nu^2[4\omega^2 + \nu(\nu+4\gamma)[\nu+2\gamma - a^2\nu(1-2q)]] \tag{252}$$

From equations (251) and (252) we see that the stationary regime (no energetic instability) is possible only if $a^2 < \min\{1, a_{cr}^2\}$, where a_{cr}^2 is a positive root of the quadratic equation for a^2 placed in the denominator of Eq. (251). The system remains stable for the fast noise ($\nu \to \infty$) and adiabatic noise ($\nu \to 0$) limits, where

$$\langle x^2 \rangle = \frac{D[1 - a^2(1 - 2q)]}{2} \tag{253}$$

The intensive numerical analysis performed in [91] supports the existence of an energetic instability generated by multiplicative trichotomous noise. The stability-nonstability transition is found to be reentrant, i.e., if the damping coefficient is lower than certain threshold value, the energetic instability appears above a critical value of the noise correlation time, but disappears again through a reentrant transition to the energetically stable state at a higher value of the noise correlation time ("orderly influence of noise").

2.16 Different stochastic models

So far we considered a stochastic oscillator with a dichotomous mass, randomly jumping between $\pm\sigma$ values. Depending on the physical application, different mathematical models are called for. For example, in the case of a random inductance on LRC circuit, the charge and current are conserved at the moment of the fluctuation, whereas for an oscillator the angle and angular momentum have to be conserved. These two cases lead to two different mathematical models,

$$m_{\pm}\frac{d^2x}{dt^2} + \gamma\frac{dx}{dt} + kx = \eta(t) \tag{254}$$

and

$$\frac{d}{dt}(m_{\pm}x) + \gamma\frac{dx}{dt} + kx = \eta(t) \tag{255}$$

So far we used the Langevin equations (254) and (255). An alternative method, which we consider in this section, is the Fokker-Planck formalism [92]. For dichotomous multiplicative noise, it is convenient to consider two coupled Fokker-Planck equation for $P_{\pm}(x, v, t)$, describing the distribution function for each of these two states. For a pendulum, described by the equation analogous to Eq. (89), such an analysis is performed in Section 1.10.2. As a result, we found cumbersome formula for the second moment, $\langle x^2 \rangle$, which becomes negative, showing the energetic instability for the strength of multiplicative noise σ^2 larger than σ_c^2, where

$$\sigma_c^2 = \frac{2\gamma(2\gamma + m\lambda(4\gamma\lambda + 4k + m\lambda^2))}{km^2\lambda^2} \tag{256}$$

It is clear that σ_c^2 is a monotonically decreasing function of m, approaching $2\gamma\lambda/k$ as m becomes large. Since $\sigma^2 < 1$, there is clearly no instability for $2\gamma\lambda > k$. If $2\gamma\lambda < k$, there is an instability for m greater than some critical value. However, for $m < \gamma[\lambda(\sqrt{1 + 16r + 32r^2} - 3 - 4r])(1 - r)^{-1}$, where $r = k/2\lambda\gamma < 1$, there is no instability for any $\sigma^2 < 1$.

For Eq. (255), describing the case of momentum conserved at the transition, the expression of the second moment is different [92], but the condition of instability has the same limit (256) as for the conserved velocity model. The same is true for the instability conditions for the higher-order moments.

There is yet a third possible model. If the change in the mass term represents actual changes in the mass due to accretion or desorption of particles by the Brownian particle, then according to Newton's third law, the addition of mass conserves momentum, whereas the loss of mass conserves velocity. For this case, the analysis of the Fokker-Planck equation shows [92] that it is no instability for any $m_\pm > 0$. Another new phenomenon is found in the numerical calculation of the distribution function $P(E)$ of energy $E = (kx^2 + mv^2)/2$, which changes with time. It turns out that this distribution function is characterized by a power-law tail that is cut off beyond some value of E, a value that increases with time. These anomalous statistics for E imply that the system has an intermittent, burst-like character: the system is quiescent for a while, then after some random wait, undergoes a large fluctuation, and then returns to its quiescent state. The higher-order stationary moments are unstable for a wider range of parameters. The limiting value of noise intensity for the n-th moment is

$$(\sigma_c^n)^2(m \to \infty) = \frac{8\gamma\lambda}{(n + 2)k} \tag{257}$$

2.17 Probability analysis

In the previous sections we considered the first two moments for some types of noise. These moments become unstable for special values of the parameters. However, linear stability analysis, based on Eq. (163), leads to a instability threshold, which is different for different moments. Therefore, in contrast to deterministic systems, the moments cannot adequately describe the global stability. The rigorous mathematical analysis of random dynamic systems shows [85] that, similar to the order–deterministic chaos transition in non-linear deterministic equations, the stability of a stochastic differential equation is defined by the sign of Lyapunov exponents. Comprehensive analysis of Lyapunov exponents in stochastic problems has been carried out [93]. Hence, one has to turn to the properties of statistical ensembles which are described by the Fokker-Planck equation associated with Eq. (163).

For the case of rapid oscillations, $\omega \gg \gamma$, one can use the procedure detailed in [94]. Replacing $x = A\sin(\omega t + \phi)$ and $y = A\omega\cos(\omega t + \phi)$ by the amplitude $A(t) \equiv \exp[u(t)]$ and phase $\psi(t) = \omega t + \phi(t)$, yields

$$x(t) = A(t)\sin\psi(t); \quad y(t) = \omega A(t)\cos\psi(t), \qquad (258)$$

For small fluctuations of mass ($\sigma^2 \ll 1$), this leads to the following equations for $u(t)$ and $\phi(t)$

$$\frac{d}{dt}u(t) = -2\gamma\cos^2\psi(t) + \xi(t)\left[2\gamma\cos^2\psi(t) + \frac{\omega}{2}\sin(2\psi(t))\right]$$

$$\frac{d}{dt}\phi(t) = \gamma\sin(2\psi(t)) - \xi(t)[\gamma\sin(2\psi(t)) + \omega\sin^2\psi(t)] \qquad (259)$$

with initial conditions $u(0) = u_0$ and $\phi(0) = \phi_0$. The assumption of rapid oscillations allows one to replace the periodic functions in Eqs. (259) and in the appropriate Fokker-Planck equation by their average value

$$\overline{\sin^2\psi} = \overline{\cos^2\psi} = \frac{1}{2}, \quad \overline{\sin^4\psi} = \overline{\cos^4\psi} = \frac{3}{8}, \quad \overline{\cos 2\psi\sin^2\psi} = -\frac{1}{4}$$
$$(260)$$

Repeating the procedure described in [94], one arrives at the Fokker-Planck equation for small ($\sigma^2 \ll 1$) and fast ($\omega \gg \gamma$) fluctuations of mass,

$$
\frac{\partial}{\partial t} P = \gamma \frac{\partial}{\partial u} P - D \left(\gamma^2 + \frac{\omega^2}{4} \right) \frac{\partial}{\partial u} P
$$

$$
+ D \left[\left(\frac{3}{2}\gamma^2 + \frac{\omega^2}{8} \right) \frac{\partial^2}{\partial u^2} P - \frac{3\gamma\omega}{2} \frac{\partial^2}{\partial u \partial \phi} P \right.
$$

$$
\left. + \left(\frac{\gamma^2}{2} + \frac{3\omega^2}{8} \right) \frac{\partial^2}{\partial \phi^2} P \right]
\tag{261}
$$

Equation (261) yields the statistical properties of $u(t)$ and $\phi(t)$

$$
\langle u(t) \rangle = u_0 - \left(\gamma - D \frac{4\gamma^2 + \omega^2}{4} \right) t;
$$

$$
\langle \phi(t) \rangle = \phi_0; \quad \langle u(t)\phi(t) \rangle = -D \frac{3\gamma\omega t}{2}
\tag{262}
$$

$$
\langle [u(t) - u_0]^2 \rangle = \frac{D\omega^2}{4} t, \quad \langle [\phi(t) - \phi_0]^2 \rangle = D \frac{3\omega^2 t}{4}
\tag{263}
$$

Hence

$$
A(t) = A_0 \exp \left\{ -\gamma \left(1 - \frac{D}{D_{cr}} \right) t \right\}
\tag{264}
$$

where

$$
D_{cr} = \frac{4\gamma}{4\gamma + \omega^2}
\tag{265}
$$

and the boundary of stability is defined such that the motion is stable for the noise strength $D < D_{cr}$, and unstable for $D > D_{cr}$. Under this condition, all the moments

$$
\langle A^n(t) \rangle = \langle e^{nu(t)} \rangle
$$

$$
= A_0^n \exp \left\{ -n \left(\gamma - D \frac{4\gamma^2 + \omega^2}{4} \right) t + n^2 D \frac{12\gamma^2 + \omega^2}{16} t \right\}
\tag{266}
$$

grow exponentially in time starting from some value of n.

2.18 Diffusion of clusters with random mass

Diffusive behavior is typical for many physical, chemical, biological and social systems. The usual approach for the dynamic description of these systems is based on the Langevin equation. The typical example is Brownian motion, considered in Section 2.1.3, where the force acting on the particle consists of a friction constant and additive white noise. Such a system exhibits normal diffusion, where an averaged squared displacement is proportional to time. However, many systems exhibit anomalous diffusion. As an example of diffusion processes in complex systems, one can mention transport processes of clusters or aggregates in biological systems, where both aggregation of clusters and random walk of their center of mass is conserved. These clusters may collide with each other and create new aggregates. One of the models which described these phenomena is based on the standard three-dimensional diffusion equation with a time-dependent diffusion coefficient [95]. The assumption of the deterministic and linear in time increase of mass M was used, $M(t) = M_0(1 + \eta t)$, where M_0 is the initial mass and $\eta > 0$ is constant. It was shown that the diffusion process is anomalous.

An another model which we have considered in the framework of the Langevin equation, where the mass of the system grows randomly due to random attachment and detachment of particles [96], and this process is described by the diffusion equation [96]. This process can also be described by stochastic step functions, such as Poisson and birth- and death-processes. We here discuss both of these approaches.

A fractal cluster of dimension d_f which diffuses in dilute solution can be described by a standard three-dimensional diffusion equation,

$$\frac{\partial p(r,t)}{\partial t} = D\Delta p(r,t); \quad p(r,0) = p(r) \qquad (267)$$

where $p(r,t)$ is the probability density of finding a cluster mass centered at r at time t, and $p(r)$ is the initial probability distribution of the cluster. The diffusion coefficient D depends on the structure of the cluster (mass, shape, etc.). As in the general theory of colloids, we assume that the diffusion coefficient varies as the inverse of the mass,

$$D = \frac{D_0}{M^\beta}, \tag{268}$$

where the exponent β takes into account the effects of cluster geometry. Assuming that the aggregates are self-similar fractals, the coefficient β is an inverse fractal dimension d_f.

The mass M of the cluster changes in time due to the attachment and detachment of particles. The process of the mass changing is random in time and it may be modelled [96] by the random step function $N(t)$ as

$$M = M(t) = M_0[1 + N(t)] \tag{269}$$

where M_0 is the initial mass of the cluster. Then, the diffusion coefficient (268) is a random function of time

$$D = D(t) = \frac{D_0}{M_0^\beta[1 + N(t)]^\beta} \tag{270}$$

Equations (267)–(270) become a stochastic partial differential equation. Then, the probability $P(r,t)$ of the cluster mass center is an averaged solution of Eq. (267),

$$P(r,t) = \langle p(r,t) \rangle^N, \tag{271}$$

where the superscript N indicates an average over all realization of the process $N(t)$. If the initial state of the system does not depend upon the process $N(0)$, one obtains from Eqs. (267) and (271),

$$P(r,t) = \int_{-\infty}^{\infty} \langle G(r - r_0) \rangle^N p(r_0) d^3 r_0 \tag{272}$$

where

$$G(r,t) = [4\pi F(t)]^{-3/2} \exp[-r^2/4F(t)]$$

$$F(t) = \int_0^t ds D(s) = \frac{D_0}{M_0^\beta} \int_0^t ds[1 + N(s)]^{-\beta} \tag{273}$$

From Eqs. (273), one can find the mean square displacement $\langle r^2(t) \rangle$ of the aggregate mass center

$$\langle r^2(t) \rangle = \langle r^2(0) \rangle + 6 \langle F(t) \rangle \qquad (274)$$

and

$$\langle F(t) \rangle = \frac{6D_0}{M_0^\beta} \int_0^t ds \sum_{k=0}^\infty (1+k)^{-\beta} P_k(s) \qquad (275)$$

where

$$P_k(t) = prob|N(t) = k| \qquad (276)$$

is the probability that at time t the process $n = N(t)$ takes the value k.

For the Poisson growth process, the probability that k-unit particles that attach to the cluster during the time interval $(0, t)$ is given by the Poisson distribution,

$$P_k(t) = \frac{(\lambda t)^k}{k!} \exp(-\lambda t) \qquad (277)$$

where λ is the mean number of unit particles that attached to the cluster per unit time, whereas for the linear birth and death grows model, the probability $P_k(t)$ in Eq. (276) is the solution of the master equation

$$dP_0(t)/dt = \mu_1 P_1(t)$$

$$dP_k(t)/dt = \lambda_{k-1} P_{k-1}(t) - \lambda_{k+\mu_k} P_k(t) + \mu_{k+1} P_{k+1}(t);$$

$$k = 1, 2, 3, \ldots \qquad (278)$$

with $\lambda_k = \lambda k$, $\mu_k = \mu k$, where λ and μ are the probabilistic rates per individual for attachment and detachment of particles to and from the aggregate, respectively. A detailed analysis of these two cases have been performed [96]. For the Poisson growth model, the mean number of unit particle attached to the cluster per unit time increases linearly with time. Different results has been obtained for the birth and death growth model, which depends on the relation between the birth transition coefficient λ and the death transition coefficient μ.

2.19 Force-free oscillator

As our last example, we consider an oscillator with a quadratic random mass without an external force,

$$m(1 + \sigma^2 + \Delta\xi)\frac{d^2x}{dt^2} + 2\gamma\frac{dx}{dt} + \omega^2 x = 0 \tag{279}$$

Analogous to the calculations performed earlier, transforming the differential equation (279) into an integro-differential equation, we apply the operator L^{-1} to Eq. (279), which gives [88]

$$x = -L^{-1}\left(\Delta\xi\frac{d^2x}{dt^2}\right)$$

$$= -\frac{\Delta}{\omega_1(1+\sigma^2)}\int \exp\left[-\frac{\gamma}{(1+\sigma^2)}(t-u)\right]$$

$$\times \sin[\omega_1(t-u)]\xi(u)\frac{d^2x}{dt^2}(u)du \tag{280}$$

and

$$\Delta\xi(t)\frac{d^2x}{dt^2} = -\frac{\Delta^2}{\omega_1(1+\sigma^2)}\int_0^t \exp\left[-\frac{\gamma}{(1+\sigma^2)}(t-u)\right]$$

$$\times \left\{\left[\frac{\gamma^2}{(1+\sigma^2)^2} - \omega_1^2\right]\sin[\omega_1(t-u)]\right.$$

$$\left. - \frac{2\gamma\omega_1}{1+\sigma^2}\cos[\omega_1(t-u)]\right\}\xi(t)\xi(u)\frac{d^2x}{dt^2}(u)du \tag{281}$$

where $\omega_1 = [\omega^2(1+\sigma^2) - \gamma^2]/(1+\sigma^2)$

Upon averaging Eq. (281) for Ornstein-Uhlenbeck noise, one can use the simplest version of the splitting of averages [88],

$$\left\langle\xi(t)\xi(u)\frac{d^2x}{dt^2}(u)\right\rangle = \langle\xi(t)\xi(u)\rangle\left\langle\frac{d^2x}{dt^2}(u)\right\rangle \tag{282}$$

Substituting (282) into linearized equation (281) and averaging yields,

$$\left[(1+\sigma^2)\frac{d^2}{dt^2} + 2\gamma\frac{d}{dt} + \omega^2\right]\langle x\rangle$$

$$-\frac{\Delta^2}{\omega_1(1+\sigma^2)}\int_0^t \exp\left[-\frac{\gamma}{(1+\sigma^2)}(t-u)\right]$$

$$\times\left\{\left[\omega_1^2 - \frac{\gamma^2}{(1+\sigma^2)^2}\right]\sin[\omega_1(t-u)] + \frac{2\gamma\omega_1}{1+\sigma^2}\cos[\omega_1(t-u)]\right\}$$

$$\times \langle\xi(t)\xi(u)\rangle\left\langle\frac{d^2x}{dt^2}(u)\right\rangle du = 0 \tag{283}$$

An analogous calculation has been performed for trichotomous noise [97].

For white noise, $\langle\xi(t)\xi(u)\rangle = \sigma^2\delta(t-u)$,

$$\left[(1+\sigma^2)\frac{d^2}{dt^2} + 2\gamma\frac{d}{dt} + \omega^2\right]\langle x\rangle = 0 \tag{284}$$

It follows from Eq. (284) that the presence of quadratic white noise of the oscillator mass leads to the increase of the oscillator mass by $1+\sigma^2$.

On the other hand, for dichotomous noise, one gets

$$\left[(1+\sigma^2)\frac{d^2}{dt^2} + 2\gamma\frac{d}{dt} + \omega^2\right]\langle x\rangle$$

$$+\frac{\Delta^2\sigma^2}{\omega_1(1+\sigma^2)}\int \exp\left[-\left(\lambda + \frac{\gamma}{(1+\sigma^2)}\right)(t-u)\right]$$

$$\times\left\{\left(\omega_1^2 - \frac{\gamma^2}{(1+\sigma^2)^2}\right)\sin[\omega_1(t-u)] + \frac{\gamma\omega_1}{(1+\sigma^2)}\cos[\omega_1(t-u)]\right\}$$

$$\times\left\langle\frac{d^2x}{dt^2}(u)\right\rangle du = 0 \tag{285}$$

Taking the Laplace transform

$$X(p) = \int_0^\infty \langle x(t)\rangle \exp(-pt)dt \tag{286}$$

of Eq. (285) yields

$$[(1 + \sigma^2)p^2 + 2\gamma p + \omega^2]X(p)$$

$$- \left\{ \frac{\Delta^2\sigma^2}{(1 + \sigma^2)^2} \frac{\omega^2 + 2\gamma(p + \lambda)}{(p + \lambda)^2(1 + \sigma^2) + 2\gamma(p + \lambda) + \omega^2} \right\}$$

$$\times \left[p^2X(p) - px(0) - \frac{dx}{dt}(0) \right] = 0 \tag{287}$$

In the absence of a driving force and for initial conditions, $x(t = 0)$ $= 0$, the mean solution $\langle x \rangle$ should relax to zero, which means that the solution of the fourth-order polynomial in p in Eq. (287) must have no roots with a positive real part. According to the Routh-Hurwitz theorem [89], this solution is obeyed for the fourth order equation $\sum_{i=0}^{i=4} a_i p^i = 0$ if the following relations hold,

$$\text{all } a_i > 0, \quad a_1 a_4 < a_2 a_3, \quad a_0 a_3^2 < a_1 a_2 a_3 - a_1^2 a_4 \tag{288}$$

where, according to Eq. (287),

$$a_0 = \omega^2 w; \quad a_1 = \omega^2 v + 2\gamma w;$$

$$a_2 = (\omega^2 + w)u + 2\gamma v - \frac{\Delta^2\sigma^2}{u^2}(\omega^2 + \lambda\gamma)$$

$$a_3 = u(\gamma + v) - \frac{2\gamma\Delta^2\sigma^2}{u^2}$$

$$a_4 = u^2; \tag{289}$$

with

$$u = 1 + \sigma^2; \quad v = 2\gamma + 2\lambda u; \quad w = \omega^2 + 2\gamma\lambda + u\lambda^2 \tag{290}$$

Equations (288) define the stability conditions in the form of the relations between two internal parameters, ω and γ, and three external parameters, σ, δ and λ.

2.20 Stochastic resonance in the oscillator with a random mass

In a broad sense, stochastic resonance means the non-monotonic dependence of the output signal as a function of some characteristic of noise or of the periodic signal. Stochastic resonance is a phenomenon found in a dynamic nonlinear system driven by a combination of a random and a periodic force [98, 99]. However, it was shown [90, 100] that stochastic resonance also occurs in a linear system subject to multiplicative color noise. The system described by Eq. (95) falls into this category. Inserting $a\sin(\Omega t)$ into (95) yields

$$[1 + \sigma^2 + \Delta\xi(t)]\frac{d^2x}{dt^2} + \gamma\frac{dx}{dt} + \omega^2 x = \eta + a\sin(\Omega t) \qquad (291)$$

Rewriting this equation as two first-order differential equations

$$\frac{dx}{dt} = y; \quad [1 + \sigma^2 + \Delta\xi(t)]\frac{dy}{dt} + \gamma y + \omega^2 x = \eta + a\sin(\Omega t) \qquad (292)$$

and averaging over the random noise, one obtains

$$\frac{d\langle x\rangle}{dt} = \langle y\rangle;$$

$$\frac{d\langle y\rangle}{dt} = \left[(1 + \sigma^2)\frac{d}{dt} + \gamma\right]\langle y\rangle$$

$$+ \Delta\left(\frac{d}{dt} + \lambda\right)\langle \xi y\rangle + \omega^2\langle x\rangle = a\sin(\Omega t) \qquad (293)$$

To split the correlator $(\xi dy/dt)$ we use the well-known Shapiro-Loginov procedure [82] which yields, for exponentially correlated noise

$$\left\langle \xi\frac{dg}{dt}\right\rangle = \left(\frac{d}{dt} + \lambda\right)\langle \xi g\rangle \qquad (294)$$

Multiplying Eqs. (292) by $\xi(t)$ and averaging yields

$$\left(\frac{d}{dt} + \lambda\right)\langle \xi x\rangle = \langle \xi y\rangle;$$

$$\left[(1 + \sigma^2 + \Delta^2)\left(\frac{d}{dt} + \lambda\right) + \gamma\right]\langle \xi y\rangle + \Delta\sigma^2\frac{d}{dt}\langle y\rangle + \omega^2\langle \xi x\rangle = \langle \xi\eta\rangle$$

$$(295)$$

For non-correlated noises, $\langle \xi\eta\rangle = 0$, one obtains the following four equations for the four functions, $\langle x\rangle$, $\langle y\rangle$, $\langle \xi x\rangle$ and $\langle \xi y\rangle$,

$$\frac{d}{dt}\langle x\rangle = \langle y\rangle;$$

$$\left[(1+\sigma)\frac{d}{dt}+\gamma\right]\langle y\rangle + \Delta\left(\frac{d}{dt}+\lambda\right)\langle \xi y\rangle + \omega^2\langle x\rangle = 0;$$

$$\left(\frac{d}{dt}+\lambda\right)\langle \xi x\rangle = \langle \xi y\rangle;$$

$$\left[(1+\sigma+\Delta^2)\left(\frac{d}{dt}+\lambda\right)+\gamma\right]\langle \xi y\rangle + \Delta\sigma\frac{d}{dt}\langle y\rangle + \omega^2\langle \xi x\rangle = 0$$

$$(296)$$

This yields a fourth-order differential equation for $\langle x\rangle$,

$$(f_1 f_2 - \Delta^2\sigma^2)\frac{d^4}{dt^4}\langle x\rangle + (\gamma f_2 + f_1 f_3 - 2\Delta^2\lambda\sigma^2)\frac{d^3}{dt^3}\langle x\rangle$$

$$+ (\omega^2 f_2 + f_1 f_4 + \gamma f_3 - \sigma^2\Delta^2\lambda^2)\frac{d^2}{dt^2}\langle x\rangle + (\gamma f_4 + \omega^2 f_3)\frac{d}{dt}\langle x\rangle$$

$$+ \omega^2 f_4\langle x\rangle = (\omega^2 f_2 + \omega^2 - f_2\Omega^2)a\sin(\Omega t)$$

$$+ (2\lambda f_2 + \gamma)a\Omega\cos(\Omega t) \qquad (297)$$

where

$$f_1 = 1+\sigma^2; \quad f_2 = 1+\sigma^2+\Delta^2;$$

$$f_3 = \gamma + 2\lambda f_2; \quad f_4 = \omega^2 + \lambda(\gamma+\lambda f_2) \qquad (298)$$

In a similar way, one can obtain the equation for the second moment $\langle x^2\rangle$ associated with Eq. (95), which is transformed into six equations for six variables, $\langle x^2\rangle$, $\langle y^2\rangle$, $\langle xy\rangle$, $\langle \xi x^2\rangle$, $\langle \xi y^2\rangle$ and $\langle \xi xy\rangle$, but we shall not write down these cumbersome equations.

Analogous to the cases of random frequency and random damping [11], we seek the solution of Eq. (297) in the form

$$\langle x\rangle = A\sin(\Omega t + \phi) \qquad (299)$$

One easily finds

$$A = \left(\frac{f_5^2 + f_6^2}{f_7^2 + f_8^2}\right)^{1/2}; \quad \phi = \tan^{-1}\left(\frac{f_5 f_7 + f_6 f_8}{f_5 f_8 - f_6 f_7}\right) \qquad (300)$$

with

$$f_5 = (f_4 - f_2\Omega^2)a; \quad f_6 = \Omega f_3 a;$$

$$f_7 = \Omega^3(\gamma f_2 + f_1 f_3 - 2\Delta^2\lambda\sigma^2) - \Omega(\gamma f_4 + \omega^2 f_3)$$

$$f_8 = \omega^2 f_4 - \Omega^2(\omega^2 f_2 + f_1 f_4 + \gamma f_3 - \Delta^2\lambda^2\sigma^2) + \Omega^4(f_1 f_2 - \Delta^2\sigma^2)$$
$$(301)$$

where the functions $f_1 \ldots f_4$ were defined in Eq. (298).

One can compare Eqs. (299)–(301) with the equations for the first moment $\langle x \rangle$, obtained for the cases of random frequency and random damping, respectively [11], subject to symmetric dichotomous noise, and extended afterwards [101], [102] to the case of asymmetric noise. All these equations are of fourth order with the same dependence on the frequency Ω of the external field, but a slightly different dependence on the noise parameters.

The amplitude A of the output signal depends on the characteristics σ, Δ, λ of the asymmetric dichotomous noise and the frequency Ω of the input signal. The signal-to-noise ratio is of frequent use in the analysis of stochastic resonance, which involves the use of the second moments. For simplicity, we call stochastic resonance the nonmonotonic behavior of the ratio A/a of the amplitude of the output signal A to the amplitude a of the input signal. (Output-Input ratio, OIR).

Figure 4 shows the dependence of the OIR on the external frequency, which confirms the existence of the phenomenon of stochastic resonance. Moreover, the presence of noise, which usually plays the destructive role, results here in an increase of the output signal, thereby improving the efficiency of a system by the amplification of a weak signal. In the absence of noise, the usual dynamic resonance occurs when the frequency of an external force approaches the eigenfrequency of an oscillator.

Figures 5 and 6 show the Ω-dependence of the OIR for parameters $\gamma = \lambda = \sigma^2 = \Delta = 1$ and different eigen-frequencies, $\omega < 1$ (Fig. 5) and $\omega > 1$ (Fig. 6). The values of the maxima increase with decreasing ω on both plots, although the positions of maxima are shifted to the right with the decreasing ω for $\omega > 1$ and to the left for $\omega < 1$. Figures 7–9 show the resonant dependence of OIR as the

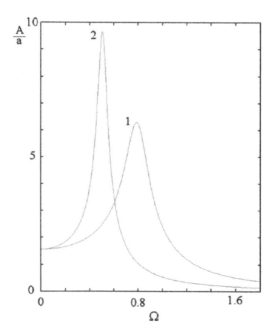

Figure 4: Stochastic resonance: output-input ratio as a function of the external frequency Ω for $\omega = 0.8$ and $\gamma = 0.2$. Curves 1 and 2 correspond to $\sigma^2 = \Delta = \lambda = 0$ and $\sigma^2 = 1.5, \Delta = 0.2, \lambda = 0.5$, respectively. Reprinted figure with permission from [28]. Copyright @ 2012, World Scientific Publication Co. Pte. Ltd.

function of the inverse correlation time λ, the strength σ^2 and the asymmetry of noise Δ.

In addition to the above considered case of an oscillator with a random mass, stochastic resonance occurs in an oscillator with two multiplicative sources of noise, described by the following dynamic equation (with $m = 1$) [103]

$$\frac{d^2x}{dt^2} + [\gamma + \xi(t)]\frac{dx}{dt} + [\omega^2 + \xi(t)]x = a\sin(\Omega t) \qquad (302)$$

where $\xi(t)$ denotes trichotomous noise such that $\xi(t) = -q, 0$ and q with correlation function

$$\langle \xi(t_1)\xi(t_2) \rangle = 2qa^2 \exp(-\lambda|t_2 - t_1|) \qquad (303)$$

The stationary ($t \to \infty$) averaged first moment can be found using the Shapiro-Loginov procedure analogous to calculations performed

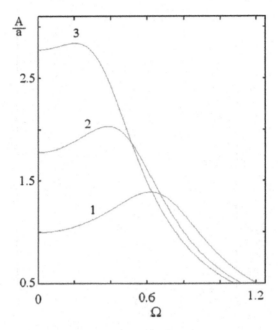

Figure 5: Output-input ratio as a function of the external frequency Ω for different $\omega \leq 1$ and $\gamma = \lambda = \sigma^2 = \Delta = 1$. Curves 1, 2 and 3 correspond to $\omega = 1$, $\omega = 0.75$ and $\omega = 0.6$ respectively. Reprinted figure with permission from [28]. Copyright © 2012, World Scientific Publication Co. Pte. Ltd.

in the previous section. One obtains [103]

$$\langle x \rangle = A \sin[\Omega t + \phi] \tag{304}$$

with A and ϕ are known functions of the oscillator parameters ω^2, γ and the noise parameters q, a, λ. The stochastic resonance ratio A/a in the following manner depends on the input signal frequency and on the noise parameters:

(1) Changes of Ω influence A/a non-monotonically. For suitably selected parameters, it displays stochastic resonance. Increase of the damping coefficient γ decreases the stochastic resonance effect.

(2) Increase of the noise switching rate λ (or decrease of the noise correlation time λ^{-1}) results in the appearance of stochastic resonance.

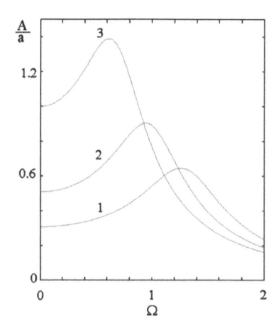

Figure 6: Output-input ratio as a function of the external frequency Ω for different $\omega > 1$ and $\gamma = \lambda = \sigma^2 = \Delta = 1$. Curves $1, 2$ and 3 correspond to $\omega = 1.8, \omega = 1.4$ and $\omega = 1.0$ respectively. Reprinted figure with permission from [28]. Copyright @ 2012, World Scientific Publication Co. Pte. Ltd.

(3) Increase of the noise strength q leads to a non-monotonic shift of the maximum of stochastic resonance curve to the right. When $q = 0.5$, the trichotomous noise degenerates into two binary sources of noise.

(4) The amplitude A of the output signal is a non-monotonic function of the input amplitude a, first showing a maximum and then a minimum, and the peak value shifts to the left.

All these results are supported by numerical simulations [103].

2.21 Stability conditions for a linear oscillator with a random mass

Here we consider the more complicated problem of the stability of the solutions. For a deterministic equation, the stability of the fixed points is given by the sign of α, which is found from the solution

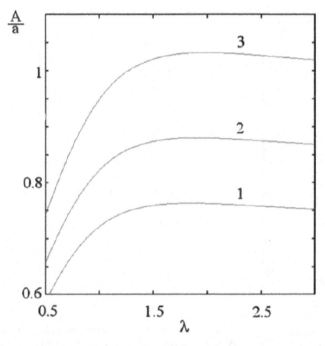

Figure 7: Output-input ratio as a function of the inverse correlation length λ for $\gamma = 0.2$, $\Omega = 1, \sigma^2 = 2.8$ and $\Delta = 2.5$. Curves 1,2 and 3 correspond to $\omega = 1.85$, $\omega = 1.8$ and $\omega = 1.75$, respectively. Reprinted figure with permission from [28]. Copyright @ 2012, World Scientific Publication Co. Pte. Ltd.

of the form $\exp(\alpha t)$ of a linearized equation near the fixed points. The situation is quite different for a stochastic equation. The first moment $\langle x(t) \rangle$ and higher moments become unstable for some values of the parameters. However, the usual linear stability analysis, which leads to instability thresholds, turns out to be different for different moments making them unsuitable for a stability analysis. A rigorous mathematical analysis of random dynamic systems shows [85] that, similar to the order–deterministic chaos transition in nonlinear deterministic equations, the stability of a stochastic differential equation is given by the sign of Lyapunov exponents λ. This means that for a stability analysis, one has to go from the Langevin-type equations to the associated Fokker-Planck equations, which describe the properties of statistical ensembles and then to calculate the Lyapunov

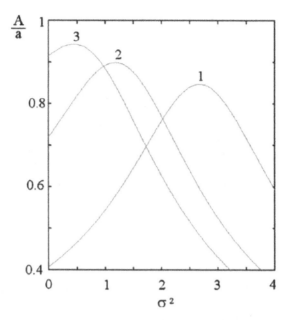

Figure 8: Output-input ratio as a function of the noise strength σ^2 for $\gamma = \lambda = \Delta = \Omega = 1$. Curves $1, 2$ and 3 correspond to $\omega = 1.8$, $\omega = 1.4$ and $\omega = 1.2$, respectively. Reprinted figure with permission from [28]. Copyright @ 2012, World Scientific Publication Co. Pte. Ltd.

index λ, defined as [85]

$$\lambda = \frac{1}{2} \left\langle \frac{\partial \ln(x^2)}{\partial t} \right\rangle = \left\langle \frac{\partial x/\partial t}{x} \right\rangle \tag{305}$$

One can see from Eq. (305) that it is convenient to replace the variable x in the Langevin equations with the variable $z = (dx/dt)/x$,

$$\frac{dz}{d\tau} = \frac{d^2 x/d\tau^2}{x} - \frac{(dx/d\tau)^2}{x^2} \equiv \frac{d^2 x/d\tau^2}{x} - z^2 \tag{306}$$

The Lyapunov index λ now takes the following form [86],

$$\lambda = \int_{-\infty}^{\infty} z P_{st}(z) dz \tag{307}$$

where $P_{st}(z)$ is the stationary solution of the Fokker-Planck equations corresponded to the Langevin equations expressed in the variable z.

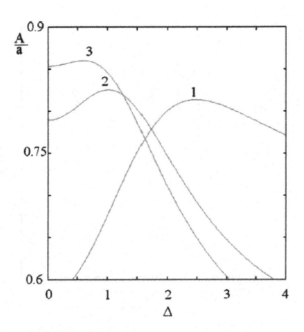

Figure 9: Output-input ratio as a function of the noise assymetry Δ for $\gamma = \lambda = \Omega = 1$. Graphs 1, 2 and 3 correspond to $\omega = 1.45$, $\omega = 1.65$ and $\omega = 1.7$, respectively. Reprinted figure with permission from [28]. Copyright © 2012, World Scientific Publication Co. Pte. Ltd.

Replacing the variable x in Eq. (95) by the variable z leads to

$$\frac{dz}{d\tau} = A(z) + \xi_1 B(z) \tag{308}$$

where

$$A(z) = -z^2 - B(z); \quad B(z) = \frac{1}{R}(1 + \sigma^2)(\gamma z + \omega^2);$$

$$\xi_1(t) = \frac{\Delta}{1 + \sigma^2}\xi(t) \tag{309}$$

2.22 White noise

As we have explained, white noise cannot be used for an oscillator with random mass. Nevertheless, this approximation might be useful for similar problems. For white noise,

$$\langle \xi_1(t_1)\xi_1(t_2) \rangle = \frac{D\Delta^2}{(1 + \sigma^2)^2}\delta(t_1 - t_2) \tag{310}$$

The Fokker-Planck equation associated with Eq. (308) has the following form (Stratonovich interpretation),

$$\frac{\partial P(z,\tau)}{\partial \tau} = -\frac{\partial}{\partial z}[A(z)P] + \frac{D\Delta^2}{2(1+\sigma^2)^2}\frac{\partial}{\partial z}B(z)\left[\frac{\partial}{\partial z}B(z)P\right]$$

(311)

For the stationary case, this equation reduces to

$$-[A(z)P_{st}] + \frac{D\Delta^2}{2(1+\sigma^2)^2}B(z)\frac{\partial}{\partial z}[B(z)P_{st}] = J \qquad (312)$$

where J is the constant probability current.

The solution of the homogeneous equation (312) (with $J = 0$) is

$$P_{st}(z) = \frac{C}{B(z)}\exp\left[\frac{2(1+\sigma^2)^2}{D\Delta^2}\int_c^z dy\,\frac{A(y)}{B^2(y)}\right] \qquad (313)$$

The solution of the inhomogeneous equation (312) can be obtained by the method of the variation of constants,

$$P_{sr}(z) = \frac{4J(1+\sigma^2)^2}{D\Delta^2 B(z)}\exp\left[\frac{2(1+\sigma^2)^2}{D\Delta^2}\int_c^z\frac{A(x)}{B^2(x)}dx\right]\int_c^z\frac{dy}{B(y)}$$

$$\times\left[\exp-\frac{2(1+\sigma^2)^2}{D\Delta^2}*\int_c^y dx\,\frac{A(x)}{B^2(x)}\right]$$

$$+\frac{C}{B(z)}\exp\left[\frac{2(1+\sigma^2)^2}{D\Delta^2}\int_c^z dy\,\frac{A(y)}{B^2(y)}\right] \qquad (314)$$

The constant C and the reference point $z = c$ are not important for our analysis, and we may set $C = 0$ for $c = -\infty$.

Inserting (309) into (314) yields the following form,

$$P_{st} = P_{sr}(w) = \frac{J\omega_1}{N\Delta\gamma_1}(w)^{-c-1}$$

$$*\exp\left[-g\left(w-\frac{1}{w}\right)\right]\int_{-\infty}^w dx\, x^{c-1}\exp\left[g\left(x-\frac{1}{x}\right)\right] \qquad (315)$$

where

$$c = \frac{4\gamma(1 + \sigma^2)^2}{D\Delta^2} \tag{316}$$

There is no need to perform an analysis of Eq. (315), because the analogous calculation has been performed for the case of random damping [86]. After substitution in Eq. (307), this yields the following result,

$$\lambda = \frac{4\int_0^\infty du K_1(8b\sinh u)\sinh[(1 - c)u]}{\pi^2[J_{\gamma/D}^2(4/D) + Y_{\gamma/D}^2(4/D)]} \tag{317}$$

where K_1 is a modified Bessel function of the second kind, and J and Y are Bessel functions of the first and second kind, respectively. The Bessel functions are always positive, and the sign of the Lyapunov exponent λ is the same as the sign of the hyperbolic function $\sinh[(1 - c)u]$, which is the sign of $1 - c$. Therefore, an oscillator with a fluctuating mass becomes unstable when $c > 1$, i.e., the instability of the fixed point $x = 0$ occurs for $D > 4(1 + \sigma^2)^2\gamma/\Delta^2$ and does not depend on the oscillator frequency ω.

It is instructive to compare these results with those obtained for additive noise and two other multiplicative sources of noise acting on the frequency and on the damping coefficient. The first moment is not changed for the case of additive noise and changed only slightly for the case of random frequency. The original frequency ω is replaced by a renormalized frequency $\sqrt{\omega^2 - \gamma^2/4}$. However, for random damping coefficient γ excited by white noise of strength D, the randomness results in the replacement of γ by $\gamma(1 - \gamma D)$. In the case of a fluctuating mass, for fast oscillations ($\omega > \gamma$), randomness leads to the renormalization of the frequency ω and (minor) renormalization of the damping coefficient γ by $\omega\sqrt{1 - \gamma D}$ and $\gamma(1 - \gamma D)$, respectively. The statistical analysis of Lyapunov exponents for fast oscillations and white noise shows that the instability occurs for large strength of noise.

Another peculiar feature of the Lyapunov index (307), which is defined by the stationary distribution function (315), is the change of its sign as a function of the oscillator and noise parameters. The integrals in (315) contain the multi-valued function x^{c-1}. Therefore, the transformed integration contour in the complex x-plane starts at $x = \infty$, makes a branch cut at $x = 0$, encircles the origin and returns to its starting point. The integrals of these type appear in the theory of Bessel functions,

$$J_\nu(w) = -\frac{\exp(i\nu\pi)}{2\pi i} \left(\frac{w}{2}\right)^\nu \int dx\, x^{\nu-1} \exp\left[-x + \frac{w^2}{x}\right] \qquad (318)$$

Finally, comparing of (318) with (307) leads to

$$\lambda \approx \frac{J_c(g)}{J_{c-1}(g)} \qquad (319)$$

If the parameter c, defined in (316), is an integer, then the oscillating behavior of the Bessel functions of integer order leads to the result that with the change of parameter g, the Lyapunov index changes its sign. This shows that with changing of the noise strength, there are many noise-induced reentrant transitions from the ordered to disordered states, and *vice versa*.

2.23 Dichotomous noise

The stationary solution of the Fokker-Planck equation, corresponding to the Langevin equation (308), has the following form [87]

$$P_{st}(z) = N \frac{B}{\sigma^2 B^2 - A^2}$$

$$\times \exp\left[-\frac{1}{2\tau} \int^z dx \left[\frac{1}{A(x) - \sigma B(x)} + \frac{1}{A(x) + \sigma B(x)}\right]\right] \qquad (320)$$

Equation (320) has been analyzed for different forms of functions $A(x)$ and $B(x)$: $A = -x$, $B = 1$ [104]; $A = x$, $B = -x$ [105]; $A = x - x^m$, $B = x$, [106]; $A = x - x^3$, $B = 1$ [107]; $A = x^3$. $B = x$ [108], [109]; $A = x - x^2$, $B = x$ [87].

The zeroes of functions $F_\pm(x) = \pm\sigma B(x) - A(x)$ determine the boundary of $P_{st}(z)$, which diverges or vanishes at the boundaries, and determine the boundary of support of $P_{st}(z)$. The latter means that a system will approach the state z located in intervals (z_2, z_1) or

(z_4, z_3), depending on its initial position. Another important characteristic of $P_{st}(z)$ is the location of its extrema, which define the macroscopic steady states. The steady states x_m of (320) obey the following equation [87]

$$
Ax_m - \frac{\sigma^2}{\lambda} B(x_m)\frac{d}{dx}A(x_m) + \frac{2}{\lambda}A(x_m)\frac{d}{dx}A(x_m) - \frac{A(x_m)^2 B(x_m)}{\lambda dB(x_m)/dx}
$$
(321)

The first term in (321) gives the deterministic steady states whereas the next term relates to the white-noise limit ($\sigma \to \infty$, $\lambda \to \infty$ with $\sigma^2/\lambda = const$). Finally, the last two terms define the corrections coming from the final λ.

For

$$
A = \alpha x^2 + \beta x + \kappa; \quad B = \beta x + \kappa
$$
(322)

with

$$
\alpha = -1; \quad \beta = -\frac{\gamma}{R}(1+\sigma^2); \quad \kappa = -\frac{\omega^2}{R}(1+\sigma^2)
$$
(323)

one obtains, according to (309),

$$
\sigma^2 B^2 - A^2 = [\alpha x^2 + (1+\sigma)(\beta x + \kappa)][-\alpha x^2 - (1-\sigma)(\beta x + \kappa)]
$$
(324)

and

$$
\frac{1}{2\tau}\int^z \frac{dx}{\alpha x^2 + (1+\sigma)(\beta x + \kappa)}
$$

$$
= -\frac{1}{2\tau\alpha(x_1 - x_2)}\left[\int^z \frac{dx}{(x - x_1)} - \int^z \frac{dx}{(x - x_2)}\right]
$$

$$
= -\frac{1}{2\tau\alpha(x_1 - x_2)}\ln\frac{z - x_1}{z - x_2}
$$
(325)

$$
\frac{1}{2\tau}\int^z \frac{dx}{-\alpha x^2 - (1-\sigma)(\beta x + \kappa)}
$$

$$
= \frac{1}{2\tau\alpha(x_3 - x_4)}\left[\int^z \frac{dx}{(x - x_3)} - \int^z \frac{dx}{(x - x_4)}\right]
$$

$$
= \frac{1}{2\tau\alpha(x_3 - x_4)}\ln\frac{z - x_3}{z - x_4}
$$
(326)

where

$$x_{1,2} = -\frac{(1+\sigma)\beta}{2\alpha} \pm \sqrt{\left(\frac{(1+\sigma)\beta}{2\alpha}\right)^2 - \frac{(1+\sigma)\kappa}{\alpha}}$$

$$= -\gamma Q_+ \pm \sqrt{\gamma^2 Q_+^2 - \omega^2 Q_+} \qquad (327)$$

$$x_{3,4} = -\frac{(1-\sigma)\beta}{2\alpha} \pm \sqrt{\left(\frac{(1-\sigma)\beta}{2\alpha}\right)^2 + \frac{(1-\sigma)\kappa}{\alpha}}$$

$$= -\gamma Q_- \pm \sqrt{\gamma^2 Q_-^2 - \omega^2 Q_-} \qquad (328)$$

with

$$Q_\pm = \frac{(1+\sigma^2)(1\pm\sigma)}{R} \qquad (329)$$

Inserting (325)–(329) into (320) gives

$$P_{st}(z) = N(z-x_1)^{-1-[2\tau\alpha(x_1-x_2)]^{-1}}(z-x_2)^{-1+[2\tau\alpha(x_1-x_2)]^{-1}}$$

$$* (z-x_3)^{-1+[2\tau(x_3-x_4)]^{-1}}(z-x_4)^{-1-[2\tau(x_3-x_4)]^{-1}} \qquad (330)$$

According to (307), the latter equation defines the boundary of stability of the fixed point $x = 0$ for different values of parameters $\gamma Q_\pm$ and $\omega^2 Q_\pm$, which depend on characteristics ω^2, γ of an oscillator and σ, Δ and τ of the noise.

2.24 Stability conditions for a nonlinear oscillator with random damping

In the previous section we analyzed the stability condition of the linear equation with a random mass. Here we consider the stability conditions for a nonlinear equation with random damping [86],

$$\frac{d^2x}{dt^2} + [r + ax^2 - \xi(t)]\frac{dx}{dt} + \alpha x + bx^3 = 0 \qquad (331)$$

where $\xi(t)$ is a white noise with intensity D.

As it was mentioned previously, the stability analysis has to be performed from the Lyapunov index, obtained from the Fokker-Planck equation for the probability density $P(z,t)$ with

$z = (dx/dt)/x$, and not via the stability of the moments from the linearized Langevin equation (331) since the higher moments always seem to be unstable. The reason has to do with the existence of long tails in the probability density of the linearized systems, which are suppressed for the nonlinear systems showing the well defined stability threshold. These long tails have a greater influence on higher moments of x which are, therefore, less and less stable.

Upon introducing dimensionless parameters

$$\gamma = \frac{r}{\sqrt{a}}; \quad \Delta = \frac{D}{\sqrt{a}}; \quad k = \frac{br}{\alpha a}, \tag{332}$$

Eq. (331) becomes

$$\frac{d^2 x}{dt^2} + \gamma[1 + x^2 - \xi(t)]\frac{dx}{dt} + x + kx^3 = 0 \tag{333}$$

Defining the new variable $z = (dx/dt)/x$, one can rewrite Eq. (333),

$$\frac{dz}{dt} + 1 + [\gamma - \xi(t)]z + z^2 = 0 \tag{334}$$

Using the Fokker-Planck equation, which corresponds to the Langevin equation (334), one can find the stationary probability $P_{st}(z)$ and the Lyapunov index Λ equal to the $P_{st}(z)$ averaged value of $\langle z \rangle$. Finally, one obtains [86]

$$\lambda = \frac{8}{C} \int_o^\infty dx K_1 \left(\frac{8 \sinh x}{\Delta} \right) \sinh \left[\left(1 - \frac{4\gamma}{\Delta} \right) x \right] \tag{335}$$

where K_1 is a modified Bessel function of the second kind and C is the normalization constant. Since the Bessel function is always positive, the sign of the Lyapunov index (positive for stable and negative for unstable states) is defined by the argument of the hyperbolic sine, i.e., the fixed point $x = 0$ is stable for $\gamma < \Delta/4$, or in the initial variables, $r < D/4$, and does not depend on the initial frequency $\alpha^{1/2}$. The numerical calculation supports these analytic results [86]. For $\gamma = 4$, the Lyapunov index first decreases for small values of the noise intensity and then increases. Such non-monotonic behavior occurs for $\gamma > 2$.

Numerical simulation shows that the temporal series of $x = x(t)$ exhibits "on-off intermittency", i.e., the small amplitude x exhibits sudden burst of activity. This phenomenon was observed previously in chaotic systems and in the systems with multiplicative noise.

2.25 Resonance phenomena

The simplest example of mechanical resonance is a non-damped harmonic oscillator subject to a periodic force, where the steady-state amplitude of the oscillator approaches infinity when the external force frequency approaches the eigen-frequency of the oscillator. This phenomenon was probably known to the ancient Egyptians who invented the water clock, but the classical demonstration of dynamic resonance are quite recent architectural flaws uncovered in the United States. The first was the Takoma bridge which was destroyed by the wind force at the resonance frequency, and the second was the Paramount Communication Building in New York where the winds twisted the top floors and pried windows loose from their casements.

The well-known phenomena of deterministic chaos, stochastic and vibrational resonances occur for an oscillator with a random mass if one adds one or two periodic forces to the oscillator equation. Stochastic resonance manifests itself in the fact that the noise, which always plays a destructive role, appears as a constructive force, increasing the output signal as a function of noise intensity. Like stochastic resonance, vibrational resonance manifests itself in the enhancement of a weak periodic signal through a high-frequency periodic field, instead of through noise as in the case of stochastic resonance.

One of great achievements of twentieth-century physics was establishing a deep relationship between deterministic and random phenomena. The widely studied phenomena of "deterministic chaos" and "stochastic resonance" might sound contradictory, consisting of both deterministic and random terms. In addition to stochastic resonance, another exciting phenomenon is deterministic chaos which appears in equations without any random force. Deterministic chaos means an exponential divergence in time of the solutions for even the smallest change in the initial conditions. Therefore, there exists a close connection between determinism and randomness, even though they are different forms of behavior [110].

Here we consider a new manifestation of the resonance of an oscillator. The dynamic equation of motion of a bistable under-damped one-dimensional oscillator driven by a multiplicative random force $\alpha\xi(t)$, an additive random force $\eta(t)$, and two periodic forces, $A\sin(\omega t)$ and $C\sin(\Omega t)$, has the following form

$$\frac{d^2x}{dt^2} + \gamma\frac{dx}{dt} - \omega_0^2 x + \alpha\xi(t)x + bx^3 = \beta\eta(t) + A\sin(\omega t) + C\sin(\Omega t)$$

(336)

The dynamic resonance mentioned above corresponds to $\gamma = b = \alpha = \beta = C = 0$ and $\omega \to \omega_0$. Let us consider some other limiting cases of Eq. (336).

(1) Brownian motion ($\omega_0 = b = A = C = 0$) has been studied most widely with many applications. The equilibrium distribution comes from the balance of two contrary processes: the random force which tends to increase the velocity of the Brownian particle and the damped force which tends to slow down the particle [90].

(2) The double-well oscillator with additive noise ($\alpha = A = C = 0$) and small damping, $\gamma \ll \omega$, exhibits two or three peaks in the power spectrum (Fourier component of the correlation function), descriptive of fluctuation transitions between the two stable points of the potential, small intra-well vibrations and the over-the-barrier vibrations [111].

(3) Stochastic resonance (SR) in overdamped ($d^2x/dt^2 = \alpha = C = 0$) and underdamped ($\alpha = C = 0$) oscillators is a very interesting and counterintuitive phenomenon, where the noise increases a weak input signal. SR occurs when the deterministic time-scale of the external periodic field is synchronized with a stochastic time-scale, determined by the Kramers transition rate over the barrier.

(4) Stochastic resonance in a linear overdamped oscillator ($d^2x/dt^2 = \beta = b = C = 0$), unlike the non-linear case, allows an exact solution [112, 113]. However, this effect occurs only when the multiplicative noise $\xi(t)$ is colored and not white.

(5) Vibrational resonance ($\alpha = \beta = 0$), which occurs in a deterministic system, manifests itself in the enhancement of a weak periodic signal through a high-frequency periodic field, instead of through noise, as in the case of stochastic resonance.

(6) "Erratic" behavior shows up as a "random-like" phenomenon in a simple system ($d^2x/dt^2 = \alpha = \beta = 0$) with two incommensurate external frequencies, ω and Ω.

2.26 Vibrational resonance

The analysis of stochastic resonance for an oscillator with fluctuating mass has been performed in Section 2.20. Like stochastic resonance, vibrational resonance manifests itself in the enhancement of a weak periodic signal through a high-frequency periodic field, instead of through noise as in the case of stochastic resonance. The deterministic equation of motion then has the following form,

$$\frac{d^2x}{dt^2} + \gamma\frac{dx}{dt} - \omega_0^2 x + \beta x^3 = A\sin(\omega t) + C\sin(\Omega t) \qquad (337)$$

Equation (337) describes an oscillator moving in a symmetric double-well potential $V(x) = -\omega_0^2 x^2/2 + \beta x^4/4$ with a maximum at $x = 0$ and two minima $x_\pm$ of depth d,

$$x_\pm = \pm\sqrt{\frac{\omega_0^2}{\beta}}; \quad d = \frac{\omega_0^4}{4\beta} \qquad (338)$$

The amplitude of the output signal as a function of the amplitude C of the high-frequency field has a bell shape, showing the phenomenon of vibrational resonance. For ω close to the frequency ω_0 of the free oscillations, there are two resonance peaks, whereas for smaller ω, there is only one resonance peak. These different results correspond to two different oscillatory processes, jumps between the two wells and oscillations inside one well.

Assuming that $\Omega \gg \omega$, resonance-like behavior ("vibrational resonance" [78]) manifests itself in the response of the system at the low-frequency ω, which depends on the amplitude C and the frequency Ω of the high-frequency signal. The latter plays a role similar to that of noise in SR. If the amplitude C is larger than the barrier height d, the field during each half-period π/Ω transfers the system from one potential well to the other. Moreover, the two frequencies ω and Ω are similar to the frequencies of the periodic signal and the Kramers rate of jumps between the two minima of the underdamped oscillator. Therefore, by choosing an appropriate relation between the input signal $A\sin(\omega t)$ and the amplitude C of the large signal (or the strength of the noise) one can obtain a non-monotonic dependence of the output signal on the amplitude C (vibration resonance) or on the noise strength (stochastic resonance). To put this another way [114], both noise in SR and the high-frequency signal in vibrational resonance change the system response to a low-frequency signal.

Let us now consider an approximate analytical solution of Eq. (337). In accordance with the two times scale in this equation, we seek a solution of the form

$$x(t) = y(t) - \frac{C\sin(\Omega t)}{\Omega^2}, \qquad (339)$$

where the first term varies significantly only over times t, while the second term varies much more rapidly. On substituting Eq. (339) into (337), one can average over a single cycle of $\sin(\Omega t)$. Then, odd powers of $\sin(\Omega t)$ vanish, whereas the $\sin^2(\Omega t)$ term gives $1/2$. This gives the following equation for $y(t)$,

$$\frac{d^2 y}{dt^2} + \gamma\frac{dy}{dt} - \left(\omega_0^2 - \frac{3bC^2}{2\Omega^4}\right)y + by^3 = A\sin\omega t \qquad (340)$$

with

$$y_{unstable} = 0; \quad y_{stable} = \pm\sqrt{\frac{\omega_0^2 - 3bC^2/2\Omega^4}{b}}; \quad d = \frac{[\omega_0^2 - 3bC^2/2\Omega^4]}{4b} \qquad (341)$$

One can say that Eq. (340) is the "coarse-grained" version (with respect to time) of Eq. (337). For $3\beta C^2/2\Omega^4 > \omega_0^2$, the phenomenon of dynamic stabilization occurs [115], namely, the high-frequency external field stablizes the previously unstable position $y_0 = 0$.

Seeking the solution of Eq. (340) near the stable points (341) of the form

$$y(t) = y_{stable} + \Theta \sin(\omega t - \theta) \tag{342}$$

and linearizing Eq. (340) in Θ gives

$$\Theta = \frac{A}{\sqrt{(\omega_1^2 - \omega^2)^2 + \gamma^2 \omega^2}} \tag{343}$$

where

$$\omega_1^2 = \frac{3bC^2}{2\Omega^4} - \omega_0^2 + 3b(y_{stable})^2 \tag{344}$$

A resonance in the linearized equation (340) occurs when $\omega_1 = \omega$, which, after substituting in Eq. (340), leads to the following relation between the amplitudes and frequencies of the two driving fields which produce the resonant behavior,

$$\omega^2 = \frac{3bC^2}{2\Omega^4} - \omega_0^2 + \frac{3bA^2}{4\gamma^2 \omega^2} \tag{345}$$

In addition to the resonance phenomenon, one can study [116] the influence of the positions and depths of the potential on the vibrational resonance. Assuming that $\omega_0^2 = b$, which means, according to Eq. (341), that the positions of the minima remain fixed, let us ask for which value of a control parameter C the ratio of the output signal Θ to the input signal A is maximal. According to Eq. (343), this occurs when $S = (\omega_1^2 - \omega^2)^2 + \gamma^2 \omega^2$ is minimal, which is determined by the condition $dS/dC = 0$, which, using (344) with $\omega_0^2 = b$, results in

$$2(\omega_1^2 - \omega^2) \frac{d\omega_1^2}{dC} = \frac{3bC}{\Omega^4} \left[\frac{3bC^2}{2\Omega^4} - b - \omega^2 + 3b(y_{stable})^2 \right] = 0 \tag{346}$$

or, for $y_{stable} = 0$,

$$C^2 = \frac{2\Omega^4}{3b}(b + \omega^2) \tag{347}$$

and for $y_{stable} = \pm\sqrt{\frac{\omega_0^2}{b} - \frac{3C^2}{2\Omega^4}}$,

$$C^2 = \frac{\Omega^4}{3b}(2b - \omega^2) \tag{348}$$

Equation (348) has real solutions for C only if $2b > \omega^2$.

Thus far, we considered equal values of two control parameters, $\omega_0^2 = b$ changing the depths of potential while keeping the positions of minima x_{stable} unaltered. Analogously, one can assume that $\omega_0^4 = b$ thereby changing the distance between the minima and not the potential depth. Then, for $y_{stable} = 0$,

$$C^2 = \frac{2\Omega^2}{3}\left(2 + \frac{\omega^2}{b}\right) \tag{349}$$

and for $y_{stable} = \pm\sqrt{\frac{\omega_0^2}{b} - \frac{3C^2}{2\Omega^4}}$

$$C^2 = \frac{2\Omega^2}{3}\left(2 - \frac{\omega^2}{b}\right) \tag{350}$$

with the proviso that $2b > \omega^2$.

The above results have been obtained for an underdamped oscillator. It turns out [117, 118] that a similar effect also takes place for an overdamped oscillator ($d^2x/dt^2 = 0$ in Eq. (337)). The influence of the additional additive noise on vibrational resonance and the advantages of vibrational resonance compared to stochastic resonance in the detection of weak signals have been studied [119].

For an oscillator with a random mass, one has to perform the preceding analysis of equation (337), based on dividing its solution in the two time scales (Eq. (339)), followed by the linearization of Eq. (340) for the slowly changing solution. The subsequent analysis of an oscillator equation with one periodic force is quite analogous to the analysis of the stochastic resonance phenomenon.

Equation (337) describes an oscillator moving in a symmetric double-well potential. The vibrational resonance in the quintic oscillator with the potential of the form

$$V(x) = -\frac{1}{2}\omega_0^2 x^2 + \frac{1}{4}bx^4 + \frac{1}{6}cx^6 \tag{351}$$

has been studied [120]. Finally, the vibrational resonance and an appearance of chaos in the Van der Pol oscillator were investigated [121]. Because of the many applications in physics, chemistry, biology and engineering, vibrational resonance still attracts great interest, and new applications will surely be found in the future.

2.27 Deterministic chaos

One of the great achievements of twentieth-century physics was the prediction of deterministic chaos which appears in equations without any random force. Deterministic chaos means an exponential increase in time of the solutions for even the smallest change in the initial conditions. Therefore, to obtain a "deterministic" solution, one would have to specify the initial conditions to an infinite number of digits. Deterministic chaos occurs if the differential equations are nonlinear and contain at least three variables. This points to the important difference between underdamped and overdamped equations of an oscillator, since deterministic chaos may occur only in the underdamped oscillator. Here, we present an example of "erratic" behavior, which, like deterministic chaos, is drawn midway between deterministic and stochastic behavior.

Consider the simple example of an overdamped oscillator subject to two periodic fields,

$$\frac{dx}{dt} + \omega^2 x = C_1 \cos(\omega_1 t) + C_2 \cos(\omega_2 t) \qquad (352)$$

We show that the solutions of this equation are "erratic", being intermediate between deterministic and chaotic.

The stationary solutions of Eq. (352) are

$$x(t) = \frac{C_1}{\omega_1} \sin(\omega_1 t) + \frac{C_2}{\omega_2} \sin(\omega_2 t) \qquad (353)$$

Replacing the continuous time in Eq. (352) by discrete times $2\pi n/\omega_2$ leads to [122]

$$x\left(n\frac{2\pi}{\omega_2}\right) = x(0) + \frac{C_1}{\omega_1} \sin\left(2\pi n\frac{\omega_1}{\omega_2}\right) \qquad (354)$$

If ω_1/ω_2 is an irrational number, the sin factor in (354) will never vanish and the motion becomes "erratic". The properties of "erratic" motion can be understood from the analysis of the correlation function associated with the n-th and $(n + m)$-th points,

$$C(2\pi m\omega_1/\omega_2) = \lim_{N\to\infty} \frac{1}{N} \sum_{n=0}^{N} x(2\pi n\omega_1/\omega_2)x[2\pi(n + m)\omega_1/\omega_2]$$

$$= x^2(0) + x(0)(C_1/\omega_1) \lim_{N\to\infty} \frac{1}{N} \sum_{n=0}^{N}$$

$$\times \{\sin(2\pi n\omega_1/\omega_2) + \sin[2\pi(n + m)\omega_1/\omega_2]\}$$

$$+ (C_1/\omega_1)^2 \lim_{N\to\infty} \frac{1}{N} \sum_{n=0}^{N} \sin(2\pi n\omega_1/\omega_2)$$

$$\times \sin[2\pi(n + m)\omega_1/\omega_2] \tag{355}$$

Using the well-known relations between the trigonometric functions, one obtains

$$C\left(m\frac{2\pi\omega_1}{\omega_2}\right) = x^2(0) + \frac{1}{2}\left(\frac{C_1}{\omega_1}\right)^2 \cos\left(m\frac{2\pi\omega_1}{\omega_2}\right) \tag{356}$$

The Fourier spectrum of the correlation function (356) depends on the ratio ω_1/ω_2. If this ratio is a rational number, the spectrum will contain a finite number of peaks. However, for irrational ω_1/ω_2, the spectrum becomes broadband, what is typical of deterministic chaos. However, this "erratic" behavior arises from a simple "integrable" equation (352), which distinguishes it from deterministic chaos.

Chapter 3

Pendulum with a Random Mass

The linear harmonic oscillator is the simplest toy model used to describe different phenomena in mechanics, optics, acoustic, electronics, engineering, etc. [123]. In this model, the net force, $-kx$, is proportional to the displacement x of the mass m from its equilibrium position $x = 0$ and directed in the opposite direction. An external periodic field $A\cos(\Omega t)$ may operate on an oscillator in two different ways. First, the periodic field may act as an external force

$$\frac{d^2 x}{dt^2} + \omega_0^2 x = \frac{A}{m}\cos(\Omega t) \tag{357}$$

where $\omega_0 = \sqrt{k/m}$ is the oscillator frequency. The solution of Eq. (357) is

$$x = C_1 \cos(\omega t + \phi) + \frac{A}{m(\omega_0^2 - \Omega^2)}\cos(\Omega t) \tag{358}$$

Second, an external force may act on the parameters of a system, say, the capacitance in a LCR circuit, which will result in a periodic change of the oscillator frequency,

$$\frac{d^2 x}{dt^2} + \left[\omega_0^2 + A_1 \cos(\Omega t)\right] x = 0 \tag{359}$$

An external force may enter the oscillator equation additively (Eq. 357) or multiplicatively (Eq. 359), yielding quite different types of solutions. In the former case, the solution of this equation has the form (358), becoming unbounded when the external frequency Ω approaching the oscillator frequency ω_0. Solutions of the Mathieu equations (359) are stable or non-stable, depending on the values of the control parameters ω_0^2, A_1 and Ω. Another difference between

(357) and (359) is the growth of the energy stored in the system, which is proportional to the amplitude, i.e., to the square root of the energy for the direct excitation (357), and proportional to the energy for the parametric excitation (359).

After considering the simplest linear model, the harmonic oscillator, it is naturally to consider the simplest non-linear model, the pendulum. The pendulum is modelled as a massless rod of length l with a point mass (bob) at its end. When the bob performs an angular deflection ϕ from the equilibrium downward position, the force of gravity mg provides a restoring torque $-mgl \sin \phi$. The Newton's second law of motion states that this torque is equal to the product of the moment of inertia ml^2 and the angular acceleration $d^2\phi/dt^2$. If we introduce damping proportional to the angular velocity, the equation of motion becomes

$$d^2\phi/dt^2 + \frac{\gamma}{ml^2} d\phi/dt + \frac{g}{l} \sin \phi = 0 \qquad (360)$$

We wish to transform the dynamic equation (360), written for zero temperature, to the stochastic equation for $T \neq 0$. In addition to an additive white noise $\eta(t)$, which describes the random collisions of the surrounding molecules with a pendulum, we assume that some of these molecules are able to adhere to the pendulum bob for some random time, thereby changing its mass from m to $m[1+\xi(t)]$. Due to the positivity of mass, the noise $\xi(t)$ cannot be white noise. The simplest assumption is symmetric dichotomous noise, which randomly jumps back and forth between $\xi(t) = \sigma$ and $\xi(t) = -\sigma$ with $\sigma < 1$ and the inverse correlation length λ. The correlation function of this noise is

$$\langle \xi(t_1)\xi(t_2) \rangle = \frac{\sigma^2}{\lambda} \exp(-\lambda|t_2 - t_1|) \qquad (361)$$

Therefore, Eq. (360) takes the following form

$$d^2\phi/dt^2 + \frac{\gamma}{m(1 + \xi)l^2} d\phi/dt + \frac{g}{l} \sin \phi = \eta(t) \qquad (362)$$

with $\langle \eta(t_1)\eta(t_2) \rangle = D\delta(T_2 - t_1)$

Multiplying Eq. (362) by $(1 + \xi)$ gives

$$(1 + \xi)d^2\phi/dt^2 + \frac{\gamma}{ml^2}d\phi/dt + (1 + \xi)\frac{g}{l}\sin\phi = \eta(t) + \eta(t)\xi(t)$$

(363)

Equation (363) corresponds to the case of a random angle.

When both angle and angular moments are conserved, Eq. (363) is replaced by a more general equation

$$\frac{d}{dt}\left[(1 + \xi)\frac{d\phi}{dt}\right] + \frac{\gamma}{ml^2}d\phi/dt + (1 + \xi)\frac{g}{l}\sin\phi = \eta(t) \qquad (364)$$

The time derivative of the fluctuation part $\xi(t)$ is given by

$$\frac{d\xi}{dt} = -\frac{D_2}{2\sigma^2}\xi + \nu(t) \qquad (365)$$

where the white noise $\nu(t)$ has the correlator

$$\langle \nu(t_1)\nu(t_2)\rangle = D_2\delta(t_1 - t_2).$$

3.1 Pendulum with a random angle

Analogously to the analysis of the oscillator with a random mass, one can rewrite the second order differential equation of the pendulum with a random mass as two first-order differential equations,

$$\frac{d\phi}{dt} = \Omega \qquad (366)$$

$$(1 + \xi)d\Omega/dt + \frac{\gamma}{ml^2}\Omega + \frac{g}{l}(1 + \xi)\sin\phi = \eta(t) \qquad (367)$$

It was assumed in Eq. (367) that the additive and multiplicative sources of noise are statistically independent, $\langle \xi(t_1)\eta(t_2)\rangle = 0$, and for non-correlated noise one can use the simplest split of correlations,

$$\langle \eta\xi\Omega\rangle = \langle \eta\xi\rangle\langle\Omega\rangle = 0 \qquad (368)$$

We briefly online the calculations.

1. Multiplying equation (366) by $\xi\Omega$ and averaging over $\xi(t)$, gives for stationary states,

$$\langle \xi\Omega^2\rangle = \lambda\langle\xi\phi\Omega\rangle = \frac{\lambda^2}{2}\langle\xi\phi^2\rangle \qquad (369)$$

2. Multiplying Eq. (367) by 2Ω, gives

$$\left[\frac{d}{dt} + \frac{2\gamma}{ml^2}\right]\Omega^2 + \left(\frac{d}{dt} + \lambda\right)\xi\Omega^2$$

$$+ \frac{2g}{l}(\Omega\sin\phi + \xi\Omega\sin\phi) = 2\eta\Omega \qquad (370)$$

and for stationary states,

$$\lambda\langle\xi\Omega^2\rangle + \frac{2\gamma}{ml^2}\langle\Omega^2\rangle + \frac{2g}{l}\left[\langle\Omega\sin\phi\rangle + \langle\xi\Omega\sin\phi\rangle\right] = 2\lambda\langle\eta\phi\rangle$$

$$(371)$$

3. Multiplying Eq. (367) by ϕ gives for stationary states

$$\langle\Omega^2\rangle - \langle\xi\Omega^2\rangle + \lambda\langle\xi\phi\Omega\rangle + \frac{g}{l}\left[\langle\phi\sin\phi\rangle + \langle\xi\phi\sin\phi\rangle\right] = \langle\eta\phi\rangle \qquad (372)$$

where we have used $\phi\frac{d\Omega}{dt} = \frac{d}{dt}(\phi\Omega) - \Omega^2$.

4. Multiplying Eq. (366) by $\sin\phi$ gives

$$\sin\phi\frac{d\phi}{dt} = -\frac{d}{dt}\cos\phi = \Omega\sin\phi \qquad (373)$$

which, after averaging for stationary states reduces to

$$\langle\Omega\sin\phi\rangle = 0 \qquad (374)$$

5. Multiplying Eqs. (367) by $2\xi\Omega$ and $\phi\xi$, respectively, one obtains, after averaging, for stationary states,

$$\left(\lambda + \frac{2\gamma}{ml^2}\right)\langle\xi\Omega^2\rangle + \frac{2g}{l}\langle\xi\Omega\sin\phi\rangle = 0 \qquad (375)$$

$$\langle\xi\Omega^2\rangle + \sigma^2\langle\Omega^2\rangle - \frac{\gamma}{ml^2}\langle\xi\phi\Omega\rangle - \frac{g}{l}\langle\xi\phi\sin\phi\rangle - \frac{g}{l}\sigma^2\langle\phi\sin\phi\rangle = 0$$

$$(376)$$

By this means, we obtain five equations, (369), (371), (372), (375) and (376), for seven variables $\langle\Omega^2\rangle$, $\langle\xi\phi\Omega\rangle$, $\langle\xi\phi^2\rangle$, $\langle\xi\Omega^2\rangle$. $\langle\phi\sin\phi\rangle$, $\langle\xi\phi\sin\phi\rangle$ and $\langle\xi\Omega\sin\phi\rangle$.

3.2 Stationary states of a pendulum

Multiplying equation (51) by $d\phi/dt$ gives

$$\frac{1}{2}\frac{d}{dt}\left(\frac{d\phi}{dt}\right)^2 - 2\omega^2\frac{d}{dt}\left[\sin^2\left(\frac{\phi_0}{2}\right) - \sin^2\left(\frac{\phi}{2}\right)\right]$$

$$+ \gamma\left(\frac{d\phi}{dt}\right)^2 = \frac{d\phi}{dt}\eta(t) \tag{377}$$

The following boundary condition: $\phi = \phi_0$, $d\phi/dt = 0$, have been used in Eq. (377).

Equation (377) contains two different functions, $d\phi/dt$ and $\sin(\frac{\phi}{2})$. One may obtain a second connection between these two functions. Using

$$\frac{d}{dt}\sin\left(\frac{\phi}{2}\right) = \frac{1}{2}\left(\frac{d\phi}{dt}\right)\cos\frac{\phi}{2} \tag{378}$$

one gets

$$\left(\frac{d\phi}{dt}\right)^2 = \frac{4}{\left[1 - \sin^2\frac{\phi}{2}\right]}\left[\frac{d\sin(\phi/2)}{dt}\right]^2 \tag{379}$$

Substituting (379) into (377) yields

$$\left(\frac{1}{2}\frac{d}{dt} + \gamma\right)\frac{4}{\left[1 - \sin^2\frac{\phi}{2}\right]}\left[\frac{d\sin(\phi/2)}{dt}\right]^2$$

$$- 2\omega^2\frac{d}{dt}\left[\sin^2\left(\frac{\phi_0}{2}\right) - \sin^2\left(\frac{\phi}{2}\right)\right]$$

$$= \left(\frac{2}{\left(1 - \sin^2\frac{\phi}{2}\right)^{1/2}}\right)\frac{d\sin(\phi/2)}{dt}\eta(t) \tag{380}$$

For $\gamma = 0$ and $\eta = 0$, Eq. (380) reduces to

$$\left[\frac{d\sin(\phi/2)}{dt}\right]^2 = \omega^2\left[1 - \sin^2\frac{\phi}{2}\right]\left[\sin^2\left(\frac{\phi_0}{2}\right) - \sin^2\left(\frac{\phi}{2}\right)\right] \tag{381}$$

For $\gamma \neq 0$, writing $\sin(\phi/2) \equiv u$ yields

$$2\frac{d}{dt}\left[\frac{1}{1-u^2}\left(\frac{du}{dt}\right)^2\right] + 4\gamma\left(\frac{du}{dt}\right)^2 + 4\omega^2 u\left(\frac{du}{dt}\right)$$

$$= 2(1-u^2)^{-1/2}\left(\frac{du}{dt}\right)\eta(t) \tag{382}$$

or

$$\frac{u}{(1-u^2)}\left(\frac{du}{dt}\right)^2 + \left(\frac{d^2u}{dt^2}\right) + \gamma(1-u^2)\left(\frac{du}{dt}\right)$$

$$+ \omega^2(1-u^2)u = 2(1-u^2)^{1/2}\eta(t) \tag{383}$$

and

$$\frac{du}{dt} = 0 \tag{384}$$

In the absence of noise, $\eta = 0$, Eq. (383) has two solutions $u = 0$ and $u = 1$ for stationary states $(d/dt... = 0)$, which corresponds the down and up positions of a pendulum ($\phi = 0$ and $\phi = 180°$). Equation (383) has to be solved numerically for different values of pendulum parameters ω^2 and γ. Equation (383) is quite complicated. However, if we are interested in the long-time limit, in particular, in the stationary states where all time derivatives vanish, one may obtain the solution in the following simple way. The time dependence of u comes from two effects: the solution of the homogeneous equation, ($\eta = 0$), and the solution of the inhomogeneous equation, ($\eta(t) \neq 0$), $u = u[t, \eta(t)]$. However, as was shown above, the two solutions of the homogeneous equation correspond to the up and down static positions of the pendulum, whereas the dependence on η occurs in the stationary states, for which all time derivatives vanish. This dependence is given by the following equation,

$$\omega^2(1-u^2)u = 2(1-u^2)^{1/2}\eta(t) \tag{385}$$

or

$$u^2 = \frac{1}{2} \pm \left[\frac{1}{4} - \frac{4}{\omega^2}\eta^2(t)\right]^{1/2}, \tag{386}$$

which, after averaging, gives

$$\langle u^2 \rangle = \frac{1}{2} \pm \left[\frac{1}{4} - \frac{4D}{\omega^2} \right]^{1/2} \tag{387}$$

Equation (387) shows that for a small intensity of noise ($D < \omega^2/16$), the pendulum will oscillate between angles defined by Eq. (387), whereas for higher intensity ($D > \omega^2/16$), there are no additional restrictions and the pendulum will perform circular motion.

3.3 Probabilistic approach to a deterministic pendulum

The motion of the classical deterministic pendulum is periodic. However, in presence of friction and periodic external force, for some values of the parameters, its motion is chaotic. This chaotic state can have the appearance of random behavior, which justifies the probabilistic approach to the deterministic pendulum model, introducing (for the deterministic problem) such definitions as "probability distribution of the angular displacement" and the "first return time" [124].

The solution of the linearized equation of the pendulum

$$\frac{d^2\phi}{dt^2} = \omega^2 \sin\phi \simeq \omega^2\phi \tag{388}$$

is $\phi(t) = \phi_0 \cos(\omega t)$. The time interval dt the pendulum spends in the interval between ϕ and $\phi + d\phi$ is given by

$$dt = \left| \frac{dt}{d\phi} \right| d\phi = \frac{d\phi}{\omega(\phi_0^2 - \phi^2)^{1/2}} \tag{389}$$

The last equation remains the ergodic hypotheses in probability theory — the amount of time the pendulum spends in a certain angular interval is proportional to the probability of the system being in that interval. One can introduce the density function $P(\phi) = [\pi(\phi_0^2 - \phi^2)^{1/2}]^{-1}$, which shows that for the oscillating pendulum bob the time spent in the upward position, where the motion is slow, is longer than the time spent in the lower vertical position. Similarly, the first return time of the pendulum initial position is represented by one delta peak for $\phi_0 = 0$ and by two delta peaks for $\phi_0 \neq 0$.

The solution of the nonlinear equation $d^2\phi/dt^2 = \omega^2 \sin\phi$ can be obtained by an iterative process, taking $\phi = A\cos(\omega t)$ as the initial solution. Substituting into the pendulum equation, one obtains

$$\frac{d^2\phi}{dt^2} = \omega^2 \sin[A\cos(\omega t)]$$

$$= -\omega^2 \left\{ A\cos(\omega t) - \frac{[A\cos(\omega t)]^3}{3!} + \frac{[A\cos(\omega t)]^5}{5!} + \cdots \right\}$$

(390)

The sinusoidal terms in the time series have odd harmonics of the fundamental frequency. For the next iteration, one may use $\phi = B\cos(\omega t) + C\cos(3\omega t)$.

For the analysis of the pendulum equation with an external periodic field, it is convenient to rewrite this equation using a time scale in units $2\pi/\omega$ and a torque in units of mgl, which transforms this equation into the following form,

$$\frac{d^2\phi}{dt^2} + \frac{1}{q}\frac{d\phi}{dt} + \sin\phi = F\cos(\omega_D t)$$

(391)

where q is the reciprocal of the damping factor, ω_D is the driving frequency, and F is the amplitude of the forcing term. Intensive numerical analysis of Eq. (391) has been performed [124] for the set of parameters (q, ω_D, F). For different values of these parameters, the pendulum motion may range from regular motion, characterized by the single frequency of the external force, to complex chaotic motion with a continuum of frequencies. For some values of F, the curve shows a single point (periodic motion with a frequency of the driving force), for $0.658 < F < 0.666$, and two points ("period doubling" with one-half of the driving frequency). For larger F, there are still a finite number of points, corresponding to periodic motion with the period equal to number of points. Finally, for sufficiently large F, the pendulum motion is chaotic, moving endlessly from one unstable periodic orbit to another. Similar considerations apply to the return time distribution [124]. The author of the last article intends to "encourage discussion of the meaning and use of probability distributions for deterministic systems".

3.4 Pendulum with random angle and random momentum

Analogously to the analysis of the oscillator with a random mass, described in the previous section, one can rewrite the second-order differential equation of the pendulum with a random mass as two first-order differential equations,

$$\frac{d\phi}{dt} = \Omega \tag{392}$$

$$(1 + \xi)\frac{d\Omega}{dt} + \left(\frac{d\xi}{dt} + \gamma\right)\Omega + \omega^2 \sin\phi = \eta(t) \tag{393}$$

The time derivative of the fluctuation part $\xi(t)$ is described by the following equation

$$\frac{d\xi}{dt} = -\frac{D_2}{2\sigma^2}\xi + \nu(t) \tag{394}$$

where the white noise $\nu(t)$ has the correlator

$$\langle\nu(t_1)\nu(t_2)\rangle = D_2\delta(t_1 - t_2).$$

To obtain the stationary second moment $\langle x^2\rangle$, we convert Eqs. (364) and (365) into the relation between correlation functions, and for splitting the correlations, we use the well-known Shapiro-Loginov procedure [82]

$$\left\langle \eta\frac{dg(\xi)}{dt} \right\rangle = \left(\frac{d}{dt} + \lambda\right)\langle\eta g(\xi)\rangle \tag{395}$$

which gives for $dg/dt = A + \eta$ and white noise $\eta(t)$ with correlator $\langle\eta(t_1)\eta(t_2)\rangle = D\delta(t_1 - t_2)$,

$$\langle\eta g\rangle = D \tag{396}$$

We use the simplest split of the correlators,

$$\langle\eta\xi y\rangle = \langle\eta\xi\rangle\langle y\rangle = 0 \tag{397}$$

for non-correlated noises $\xi(t)$ and $\eta(t)$.

We briefly outline the calculations

1. Multiplying equation (392) by $\xi\Omega$, gives

$$\xi\Omega\frac{d\phi}{dt} = \xi\Omega^2 \tag{398}$$

Averaging for stationary states $(d/dt... = 0)$ results in

$$\langle \xi \Omega^2 \rangle = \lambda \langle \xi \phi \Omega \rangle \tag{399}$$

2. Multiplying Eq. (393) by η and averaging, one gets for stationary states,

$$(\lambda + \gamma)\langle \eta \Omega \rangle + \omega^2 \langle \eta \sin \phi \rangle = D \tag{400}$$

3. Substituting Eq. (394) into (393), results in

$$(1 + \xi)\frac{d\Omega}{dt} + \left[\left(-\frac{D_2}{2\sigma^2}\xi + \nu(t) \right) + \gamma \right]\Omega + \omega^2 \sin \phi = \eta(t) \tag{401}$$

Multiplying the last equation by Ω yields

$$\frac{1}{2}\frac{d\Omega^2}{dt} = -\frac{1}{2}\xi\frac{d\Omega^2}{dt} + \frac{D_2}{2\sigma^2}\xi\Omega^2 - (\nu + \gamma)\Omega^2 - \omega^2 \left\langle \frac{d\phi}{dt}\sin \phi \right\rangle + \langle \eta \Omega \rangle \tag{402}$$

and for the averaged stationary states,

$$0 = -\frac{1}{2}\left(\lambda + \frac{D_2}{\sigma^2} \right)\langle \xi\Omega^2 \rangle - (\nu + \gamma)\langle \Omega^2 \rangle + D \tag{403}$$

4. Multiplying Eq. (394) by ϕ^2, one gets

$$\phi^2\frac{d\xi}{dt} = \frac{d}{dt}(\xi\phi^2) - 2\xi\phi = -\frac{D_2}{2\sigma^2}\xi\phi^2 + \nu\phi^2, \tag{404}$$

which for averaged stationary states reduces to

$$\langle \nu\phi^2 \rangle = \frac{D_2}{2\sigma^2}\langle \xi\phi^2 \rangle - 2\langle \xi\phi \rangle \tag{405}$$

5. Multiplying Eq. (401) by ξ yields

$$\xi\frac{d\Omega}{dt} = -\sigma^2\frac{d\Omega}{dt} + \frac{D_2}{2}\frac{d\phi}{dt} - (\nu + \gamma)\xi\Omega - \omega^2\xi\sin \phi + \xi\eta \tag{406}$$

which for averaged stationary states reduces to

$$(\gamma + \nu + \lambda)\langle \xi\Omega \rangle + \omega^2\langle \xi\sin \phi \rangle = 0 \tag{407}$$

In Eq. (407), we used the simplest split of the correlators,

$$\langle \eta \xi \rangle = \langle \eta \rangle \langle \xi \rangle = 0 \tag{408}$$

for non-correlated noise $\xi(t)$ and $\eta(t)$.

6. Multiplying Eqs. (392) and (401) by Ω and ϕ, respectively, summing and averaging the equations, one gets

$$\frac{d}{dt}(\phi\Omega) = \Omega^2 - \xi \left[\frac{d}{dt}\phi\Omega - \Omega^2 \right]$$

$$- \left[-\frac{D_2}{2\sigma^2}\xi\phi\Omega + (\nu + \gamma)\phi\frac{d\phi}{dt} \right] - \omega^2\phi\sin\phi + \eta\phi \tag{409}$$

and for the averaged stationary states,

$$\langle \Omega^2 \rangle + \langle \xi\Omega^2 \rangle + \frac{\lambda}{2}\left(\frac{D_2}{2\sigma^2} - \lambda \right)\langle \xi\phi^2 \rangle - \omega^2\langle \phi\sin\phi \rangle = 0 \tag{410}$$

7. Multiplying Eq. (401) by $\nu\Omega$, and using the Shapiro-Loginov procedure, and then averaging for stationary states yields

$$\left(\frac{\lambda}{2} + \gamma \right)\langle \nu\Omega^2 \rangle + D_2\langle \Omega^2 \rangle = 0 \tag{411}$$

8. Multiplying Eq. (401) by $2\xi\Omega$ and Eq. (409) by ξ and performing the transformations described above, one obtains for the averaged stationary states,

$$[\lambda + 2\gamma]\langle \xi\Omega^2 \rangle - \frac{D_2}{2}\langle \Omega^2 \rangle + 2\omega^2\langle \xi\Omega\sin\phi \rangle = 0 \tag{412}$$

$$-\lambda\langle \xi\phi\Omega \rangle + \langle \xi\Omega^2 \rangle + \sigma^2\langle \Omega^2 \rangle - \frac{\lambda}{2}\langle \nu\phi^2 \rangle$$

$$- \frac{\lambda}{2}(\nu + \gamma)\langle \xi\phi^2 \rangle - \omega^2\langle \xi\phi\sin\phi \rangle = 0 \tag{413}$$

By this means we obtain six equations, (399), (403), (410), (411). (412) and (413), for eight variables, $\langle \Omega^2 \rangle$, $\langle \xi\phi^2 \rangle$, $\langle \xi\Omega^2 \rangle$, $\langle \eta_1\phi^2 \rangle$, $\langle \eta_1\Omega^2 \rangle$, $\langle \phi\sin\phi \rangle$, $\langle \xi\Omega\sin\phi \rangle$ and $\langle \xi\phi\sin\phi \rangle$.

For $d\xi/dt = 0$, which means that $D_2 = \nu = 0$, the six equations reduce to those considered in [28], as required. We consider the energetic instability, which corresponds to negative second moments. The latter can be found from the system of the foregoing six equations,

$$\langle \phi^2 \rangle = \frac{D(\gamma + \lambda/2)}{\lambda \omega^2 (\lambda + \gamma + \omega^2/2)} \left[\frac{\beta_1 (\lambda + 2\gamma) - \lambda \omega^2 \alpha_1 + \beta_1 D_2}{D_2 (\alpha_1 \beta_2 - \alpha_2 \beta_1)} + \frac{1}{\gamma + \lambda/2} \right]$$

(414)

and analogously

$$\langle \Omega^2 \rangle = \frac{\gamma_1}{4} \frac{\beta_1 \lambda - 4\beta_2}{\alpha_1 \beta_2 - \alpha_2 \beta_1}$$

(415)

where

$$\alpha_1 = 1 + \frac{\sigma^2}{D_2}(\lambda + 2\gamma); \quad \beta_1 = \frac{\lambda \sigma^2 \omega^2}{D_2} - \left[\omega^2 + \frac{\lambda}{2}(\lambda + \gamma) + \frac{\lambda D_2}{4\sigma^2} \right];$$

$$\gamma_1 = \frac{D_1}{\lambda}(\lambda + 2\gamma); \quad \beta_2 = \frac{\lambda D_2 \omega^2}{4\sigma^2} + \lambda \omega^2 \left[1 - \frac{\gamma}{D_2}\left(\gamma + \frac{\lambda}{2} \right) \right];$$

$$\alpha_2 = \frac{1}{2}\left(\gamma + \frac{\lambda}{2} \right)\left(\frac{D_1}{\sigma^2} - \lambda \right) + [\lambda + \gamma]\left[1 - \frac{\gamma}{D_2}\left(\gamma + \frac{\lambda}{2} \right) \right]$$

(416)

It is evident from Eqs. (414) and (415) that for some sets of the noise parameters, σ and λ, as well as the pendulum parameters, the second moments become negative, which implies energetic instability of a system.

3.5 Josephson junction with multiplicative noise

Josephson junctions — two superconductors separated by a thin insulating or metallic barrier — are very interesting counter-intuitive physical objects. Suffice it to mention such unusual phenomena as supercurrents, the dc current of the Cooper pairs without voltage and ac current induced by a constant voltage. When a junction is subject to noise, this noise is able to increase the impinging signal (stochastic resonance). The other features of Josephson junctions, such as the application of their equations to many other fields (motion of a pendulum, Brownian motion in periodic potential, etc.) and many other applications (voltmeters, magnetometers, precise measurements of $\hbar/e$), [125] have attracted considerable attention. There are hundreds of articles devoted to different properties and applications of Josephson junctions. We consider the influence of multiplicative noise, which describes the fluctuations of external parameters on the properties of Josephson junctions, a subject which is less well understood. Josephson based his discovery on the fact that each superconductor is characterized by a two-dimensional order parameter, having amplitude Ψ_0 and phase ϕ_0 of the wave function, whereas only $|\Psi_0|^2$ has physical meaning. Two separate superconductors may have different phases, and the phase difference ϕ defines the properties of a system. In the study of the Josephson junction, one considers the voltage-current characteristic, the connection between the average potential difference

$$\langle V \rangle = \hbar \langle d\phi/dt \rangle /2e \tag{417}$$

and the external current J.

The appropriate dynamic equation has the following form [125]

$$\frac{\hbar C}{2e}\frac{d^2\phi}{dt^2} + \frac{\hbar}{2eR}\frac{d\phi}{dt} + J_0 \sin\phi = J + \xi(t) \tag{418}$$

where C, R, J_0 and J are the capacitance, the junction resistance, Josephson current J_0 and external current J. The random force $\xi(t)$ has the correlator $\langle \xi(t_1)\xi(t_2) \rangle = D\delta(|t_1 - t_2|)$. The purpose of our calculation is to determine the impact of multiplicative noises, added to Eq. (418), on the current-voltage, $J - V$ characteristics of the Josephson junction. For the similar problems, one replaces J by the coordinate x for oscillator or by the current J for resistor-inductor-capacity contour, etc., and studies the average moments of these functions.

From three possible types of multiplicative noise, only one — random Josephson current J_0 (which corresponds to random frequency for a pendulum) — has been considered in detail. We bring the main results of this analysis in the next chapter. In two following chapters we bring our results for random capacitance C and random conductivity $1/R$ (random damping and random mass for a pendulum). In conclusion, we compare possible ways to improve the efficiency of different theoretical methods.

3.5.1 *Random current*

There are many results for numerical solutions of Eq. (418) with random coefficient J_0, references to which can be found in a recent article [126]. We bring here some of these results. Transferred to the scaling quantities, Eq. (418) with an additional multiplicative noise takes the form

$$m\frac{d^2\phi}{dt^2} + \gamma\frac{d\phi}{dt} + [1 + \sigma\xi_2(t)]\sin\phi = j + \xi_1(t), \qquad (419)$$

where m and $\gamma = (\hbar m/2eJ_0CR^2)^{1/2}$ are the normalized mass and junction conductance, j is the external torque, and $\xi_1(t)$ and $\xi_2(t)$ are random functions with Ornstein-Uhlenbeck correlators,

$$\langle\xi_i(t)\xi_i(t)\rangle = E^2\exp(-\lambda|t_1 - t_2|); \quad i = 1, 2$$

$$\langle\xi_i(t)\xi_j(t)\rangle = s^2\exp(-\lambda|t_1 - t_2|); \quad i \neq j$$

$$d\xi_{1,2}/dt = -\lambda\xi_{1,2} + \lambda\eta_{1,2}$$

$$\langle\eta_i(t_1)\eta_j(t_2)\rangle = 2D\delta_{ij}\delta(t_2 - t_1) \qquad (420)$$

For the medium value of correlation rate $\lambda = 0.2$, ϕ makes periodic rotation in one direction ("running state") and the average voltage $\langle V \rangle$ is not zero, while for $\lambda = 10^2$ and $\lambda = 10^{-9}$, ϕ oscillates randomly near its own equilibrium position ("locking state") with $\langle V \rangle = 0$. The numerical solution of $\phi(t)$ has been performed for the following values of parameters $E = 0.1$, $\gamma = 10$, $\sigma = 20$, $m = 0.3$ and $s = 1$.

(a) $\langle V \rangle$ as a function of the correlation rate λ for $\sigma = -20, -10,$ $-5, 5, 10, 20$, and as a function of the strength σ of multiplicative noise for $\lambda = 10^{-4}$, 10^{-2}, 10^0. For $\sigma \neq 0$, $\langle V \rangle$ displays non-monotonic behavior with λ, and, at some λ, $\langle V \rangle$ has a maximum. Moreover, for $\sigma > 0$ and $\sigma < 0$, both ϕ and $\langle V \rangle$ turn out clockwise and counterclockwise, respectively, with $\langle V \rangle$ being negative in the first case and positive in the second one. For $\sigma = 0$, both $\langle V \rangle = 0$ and ϕ fluctuates near its equilibrium positions. In other words, multiplicative noise induces a net voltage in the Josephson junction. The other parameters are the same as in the previous case.

(b) $\langle V \rangle$ as a function of the mass m for $\lambda = 10^{-5}$, 10^{-2} and 10^0 and as a function of λ for $m = 3 * 10^1$, $3 * 10^2$, $3 * 10^3$, $3 * 10^4$ and $3 * 10^5$. For increasing m, $\langle V \rangle$ increases and becomes a two-valued function. A further increase of m again leads to a single valued function. Finally, for m approaching infinity, $\langle V \rangle$ approaches zero. Furthermore, $\langle V \rangle$ varies non-monotonically with m for small λ. For larger $\lambda(10^{-2}$ and $10^0)$, the absolute value of $\langle V \rangle$ decreases when m increases.

(c) $\langle V \rangle$ as a function of the noise correlation strength s for $\lambda = 10^{-4}$, 10^{-2}, 10^0, and as a function of λ for $s = -1, -0.7, 0, 0.7$ and 1. For $\sigma > 0$, $\langle V \rangle < 0$ and ϕ turn over clockwise as $j > 0$, while $\langle V \rangle > 0$ and ϕ turn over counterclockwise as $j < 0$. If $s = 0$, i.e., it is no correlation between the additive and multiplicative noise, $\langle V \rangle = 0$, and the random motion near the equilibrium position corresponds to the locking state.

With increasing j, the increase of $\langle V \rangle$ with λ is more and more pronounced. One of the conclusions is that the maximum increase of voltage occurs at larger values of λ and j.

The dynamic equations become slightly less cumbersome if we consider the overdamped equation of the form

$$\frac{d\phi}{dt} + [J_1 + \xi(t)] \sin \phi = J_2 + \eta(t) \tag{421}$$

where both sources of noise $\eta(t)$ and $\xi(t)$ are non-symmetric dichotomous noises with $\langle \eta \rangle = \langle \xi \rangle = 0$ and the correlators

$$\langle \eta(t_1)\eta(t_2) \rangle = A_1 B_1 \exp[-\alpha|t_2 - t_1|]$$

$$\langle \xi(t_1)\xi(t_2) \rangle = A_2 B_2 \exp[-\beta|t_2 - t_1|] \tag{422}$$

Replacing the Langevin equation (421) by the appropriate Fokker-Planck equation, leads to a quite complicate formula [127] , and here we will bring only the limiting form of $\langle d\phi/dt \rangle$ as a function of J_2 (voltage-current characteristic of the process) for some special cases.

3.5.2 *White additive noise and very fast or very slow multiplicative noise*

Consider first white additive noise, $A_1 = |B_1| \equiv A$ and $\lim(A^2/\alpha) = D$. Then, for fast multiplicative noise ($\beta \to \infty$), one obtains

$$\left\langle \frac{d\phi}{dt} \right\rangle = \frac{\sinh(\pi J_2/D)}{\pi/D} \left| I_{iJ_2/D} \frac{J_1}{D} \right|^{-2} \qquad (423)$$

where $I_{iJ_2/D}$ is the modified Bessel function of the first order with imaginary argument and the imaginary index. The $I_{iJ_2/D}$ factor in (423) decreases with D, which makes it easier for a system to overcome the potential barrier between the two states, while the sinh factor makes the system more homogeneous.

For very slow process, $\beta \to 0$, the total voltage $\langle d\phi/dt \rangle$ is given by the average of the two potentials. For both $\beta \to 0$ and $\beta \to \infty$, the current as a function of β has both a minimum and a maximum [128].

3.5.3 *White additive and multiplicative noise*

For $A_1, A_2, \alpha, \beta \to \infty$ with $\lim(A_1^2/\alpha) = D_1$ and $\lim(A_2^2/\beta) = D_2$, one obtains

$$\left\langle \frac{d\phi}{dt} \right\rangle = 2\pi D_1 \left[1 - \exp\left(\frac{2\pi J_2}{D_2} \right) \right] \left\{ \int_{-\pi}^{\pi} d\xi \exp\left[\left(\frac{1}{D_1} \right) \int_{-\pi}^{\xi} g(\rho)d\rho \right] \right.$$

$$\left. * \left[\int_{\xi}^{\xi+2\pi} \exp\left(-\frac{1}{D_1} \int_{-\pi}^{z} g(\rho)d\rho \right) dz \right] \right\}^{-1} \qquad (424)$$

where $g(\phi) = J_2 - J_1 \sin \phi$. It is interesting that for weak additive and strong multiplicative noise, say $D_1 = 0.02$ and $D_2 = 10$, the voltage-current characteristic for small J_2 exceeds the result that corresponds to the Ohm's law.

3.5.4 *Multiplicative dichotomous and multiplicative additive noise*

The numerical solution for this case shows [127] that: (1) The noise has no influence on the current-voltage characteristic for multiplicative noise $J_1 > J_2 + A_1$ and for additive noise $J_2 < J_1 - B_1$, but the joint action of both sources of noise changes the current-voltage characteristic even in these regions, (2) The "ratchet effect" (appearance of net voltage in the absence of driving current, $J_2 = 0$) exists for non-symmetric additive dichotomous noise even in the presence of symmetric dichotomous multiplicative noise.

For the cases of multiplicative dichotomous and additive white noise as well as for additive dichotomous and white multiplicative noise, numerical solutions have been obtained [128].

3.5.4.1 *Random capacitance*

We recently considered [113] an oscillator with a random mass, which describes a new type of Brownian motion — Brownian motion with adhesion, where the molecules of the surrounding medium not only randomly collide with the Brownian particle, which produces its well-known zigzag motion, but they also stick to the Brownian particle for some random time, thereby changing the mass of the Brownian particle. As a result, there are restrictions on both the size of the Brownian particle, which has to be larger than the surrounding molecules, and also on its mass.

The analogous situation may take place for Josephson junction considered here. For the electrical circuit, Eq. (419) has to be replaced by the following equation,

$$m[1 + \sigma\xi_2(t)]\frac{d^2\phi}{dt^2} + \gamma\frac{d\phi}{dt} + \sin\phi = j + \xi_1(t) \qquad (425)$$

The situation is slightly different for a random pendulum, which is also described by the equation of a Josephson junction. The position of the pendulum bob is described by the angle ϕ of the spur with respect to its vertical position. The driving moment of the bob, induced by gravity, $mgl\sin\phi$, is balanced by the rotary moment $ml^2 d\phi/dt$. If the damping is proportional to the angular velocity $d\phi/dt$, which is balanced by the external random force $\xi_1(t)$, the equation of motion of the pendulum has the following form

$$\frac{d^2\phi}{dt^2} + \frac{\gamma}{ml^2[1 + \sigma\xi_2(t)]}\frac{d\phi}{dt} + \frac{g}{l}\sin\phi = \xi_1(t) \qquad (426)$$

Multiplying Eq. (426) by $1 + \sigma\xi_2(t)$ gives

$$[1 + \sigma\xi_2(t)]\frac{d^2\phi}{dt^2} + \gamma_1\frac{d\phi}{dt} + \frac{g}{l}[1 + \sigma\xi_2(t)]\sin\phi = \xi_1(t) \qquad (427)$$

where $\gamma_1 = \gamma/\{ml^2[1 + \sigma\xi_2(t)]\}$. We have used the simplest splitting of the correlators, $\langle\xi_1(t)\xi_2(t_2)\rangle = 0$.

3.5.5 *Random damping*

To our knowledge, the random damping in a pendulum equations has been intensively studied for a small oscillations, where $\sin\phi \approx \phi$, and the non-linear pendulum problem reduces to the linear oscillator problem. For a non-linear Josephson junction (or pendulum) with random damping, the equation of motion has the following form

$$m\frac{d^2\phi}{dt^2} + \gamma[1 + \sigma\xi_2(t)]\frac{d\phi}{dt} + \sin\phi = j + \xi_1(t) \qquad (428)$$

3.6 Order and chaos: are they contradictory or complimentary?

Considerable advances were made by physics during the twentieth century, culminating in the relativity and the uncertainty principle in quantum mechanics. These are now part of mainstream thought and go far beyond the context of their original discovery. Less well understood, but not less an important achievement of modern physics is the establishment of a deep interrelationship between deterministic and random phenomena. The widely studied phenomena of "deterministic chaos" and "stochastic resonance" might sound internally contradictory, consisting of half-deterministic and half random terms. However, this impression is incorrect due to the close connection between determinism and randomness, two apparently opposite forms of behavior. This connection can be illustrated by the simple example of a pendulum.

Consider a pendulum with moment of inertia I which executes a rotary plane motion in a field. The Langevin equation of motion has the following form

$$I\frac{d^2\theta}{dt^2} + \eta\frac{d\theta}{dt} \equiv T_{\text{det}} + T_{\text{ran}} = [A\cos(\omega t) + B\cos(\Omega t)]\sin\theta + 2D\xi(t)$$

$$(429)$$

where $\xi(t)$ is the delta-correlated random torque, $\langle\xi(t_1)\xi(t_2)\rangle = \delta(t_1 - t_2)$ and T_{det} and T_{ran} are deterministic external and random torques composed of two periodic signals.

Let us consider different limiting cases of Eq. (429).

DETERMINISTIC MOTION. For large viscosity η, one can neglect the inertial term in Eq. (429). In the absence of the field $B\cos(\Omega t)$ and the random force ξ, the simplified equation

$$\eta\frac{d\theta}{dt} = A\cos(\omega t)\sin\theta \qquad (430)$$

has an exact solution

$$\tan\frac{\theta}{2} = \tan\frac{\theta_0}{2}\exp\left[\frac{A}{\omega\eta}\sec(\omega t)\right] \qquad (431)$$

where $\theta_0 = \theta\ (t = 0)$. Since the external torque vanishes at $\theta = n\pi$, a pendulum which started in the interval $m\pi < \theta_0 < (m+1)\pi$, will oscillate forever in this interval. The initial condition $\theta_0 = \theta(t = 0)$ uniquely determines the values of θ for all t, i.e. the motion is fully deterministic.

RANDOM MOTION. If one adds a random force to equation (430), the pendulum is capable of overcoming barriers at $\theta = n\pi$ and the motion becomes random (noise-induced instability).

DETERMINISTIC CHAOS. However, the pendulum along with an oscillator are the typical toy model widely used in analytical studies of deterministic chaos. As an example, for the Hamiltonian case without a random force ($B = D = 0$ in Eq. (429)).

$$I\frac{d^2\theta}{dt^2} + \eta\frac{d\theta}{dt} \equiv A\cos(\omega t)\sin\theta \qquad (432)$$

deterministic chaos occurs for some values of the parameters [129]. The dynamics of deterministic chaos closely resembles random motion and can be differentiated from it only by quite subtle criteria.

Hence the previous analysis show that the general equation (429) contain deterministic motion, deterministic chaos and random motion as special cases.

Here we want to consider yet another possibility of non-trivial behavior, intermediate between deterministic and random motion, when T_{ran} in Eq. (429) is absent and the two characteristic frequencies ω and Ω are incommensurate, i.e., their ratio is irrational. For this special case, Eq. (429) reduces to

$$\frac{d\theta}{dt} = [A\cos(\omega t) + B\cos(\Omega t)]\sin\theta \qquad (433)$$

Integrating this equation leads to

$$u(t) = \frac{A}{\omega}\sin(\omega t) + \frac{B}{\Omega}\sin(\Omega t) \qquad (434)$$

where

$$u(t) = \ln\left[\frac{\theta(t)}{2}\right] \qquad (435)$$

Let us replace the continuous time in equations (433)–(435) by the discrete time $2\pi n/\Omega$, making thereby successive "snapshots" at equally spaced time intervals (stroboscopic plot). Equation (434) then becomes

$$u(n\frac{2\pi}{\Omega}) = \frac{A}{\omega}\sin(2\pi n\frac{\omega}{\Omega}) + u(0) \qquad (436)$$

or

$$\tan\left[\frac{\theta(2\pi n/\Omega)}{2}\right] = \tan\left[\frac{\theta(t)}{2}\right]\exp\left[\frac{A}{\omega}\sin(2\pi n\frac{\omega}{\Omega})\right] \qquad (437)$$

Although we have two frequencies in our problem, ω and Ω, the result is fully determined by their ratio ω/Ω.

1. When ω/Ω is an integer, $\sin(2\pi n\frac{\omega}{\Omega})$ vanishes, and the motion becomes periodic ("locked"). Indeed, starting from some $\theta(t=0)$, a pendulum will come back to this point after some integral number of cycles.

2. When ω/Ω is a rational number, which can be represented as the ratio of two integers, $\omega/\Omega = p/q$, then $\sin(2\pi n\frac{\omega}{\Omega})$ vanishes after $n = q$ cycles, and the motion is again periodic.

3. On the other hand, if ω/Ω is irrational, the sin factor in (436) will never vanish. This results in an incommensurate increase of θ_n at each recursion stage, and the motion becomes "erratic".

Therefore, the simplest way to move the pendulum motion completely random is to add to equation (430) a random torque (the term D in Eq. (429). There are, however, two ways to destroy the deterministic motion, replacing it by a motion which is intermediate between determinism and chaos. The first possibility is to replace the overdamped motion by underdamped motion by adding the inertial term described by the second derivative. Then, the system has "one-and-half" degrees of freedom, which is sufficient for the onset of the deterministic chaos [129]. As was shown above, there is also another way to destroy the pure deterministic motion of a pendulum, replacing it by quasi-periodic motion by the use of two incommensurate external frequencies (Eq. (433)). This mostly pedagogical example shows that deterministic and random phenomena complement, rather than contradict each other.

Our analysis has been fully classic. Interestingly, the analysis of the Kronig-Penny type problem with two incommensurate space periods in quasi-lattices [130] leads to some erratic behavior (not in time, like in our case, but in space), intermediate between order and chaos. Then, for commensurate frequencies, the average energy grows quadratically in time, while for incommensurate frequencies, it grows linearly for short times and saturates for long times.

As we have seen previously, the standard equation of the damped pendulum with thermal noise $\eta(t)$ has the following form

$$m\frac{d^2\phi}{dt^2} + \frac{\gamma}{l^2}\frac{d\phi}{dt} + \omega_0^2 \sin\phi = m\eta(t) \tag{438}$$

For the pendulum with additive noise, described by Eq. (438), one can replace $\gamma \to m\gamma$ and $\omega_0^2 \to m\omega_0^2$, which allows one to set $m = 1$ by the appropriate choice of units. Then, the stability conditions do not depend on the mass of the pendulum. However, the situation is different when, along with additive noise, the appropriate equations contain external fluctuations, which are described by multiplicative noise for random frequency ω_0 and random damping γ, due to the random vertical oscillations of the suspension point or the external force and external torque, respectively. In the latter case, one cannot eliminate the mass from the equation of motion, which leads to new phenomena. In particular, a new type of pendulum motion appears when the motion of a pendulum depends on its mass. In the case of small oscillations, one can replace $\sin \phi$ by ϕ, and obtain the exact analytical solution of Eq. (438), as shown previously.

Recently [28], we considered another way of introducing multiplicative noise, namely, via a fluctuating mass term,

$$m[1 + \xi(t)]\frac{d^2\phi}{dt^2} + \frac{\gamma}{l^2}\frac{d\phi}{dt} + \omega_0^2 \sin \phi = m\left[1 + \xi(t)\right]\eta(t) \qquad (439)$$

Upon multiplying the last equation by $1 - \xi(t)$, one obtains

$$m(1 - \xi^2)\frac{d^2\phi}{dt^2} + (1 - \xi(t))\left[\frac{\gamma}{l^2}\frac{d\phi}{dt} + \omega_0^2 \sin \phi\right] = m[1 - \xi^2(t)]\eta(t)$$
$$(440)$$

which, for symmetric dichotomous noise $\xi(t) = \pm\sigma$, takes the following form

$$m(1 - \sigma^2)\frac{d^2\phi}{dt^2} + (1 - \xi(t))\left[\frac{\gamma}{l^2}\frac{d\phi}{dt} + \omega_0^2 \sin \phi\right] = m[1 - \sigma^2(t)]\eta(t)$$
$$(441)$$

Equation (440) is the dynamic equation with a mass that can both increase and decrease due to fluctuations, but always remaining positive. The latter condition requires some additional constraint on the random force. One possibility is to consider the positive random force $\xi^2(t)$, which represents the fact that the mass of the oscillator may only increase due to the adhesion of the molecules of the surrounded media,

$$[1 + \xi^2(t)]\frac{d^2\phi}{dt^2} + \frac{\gamma}{l^2}\frac{d\phi}{dt} + \omega_0^2 \sin \phi = [1 + \xi^2(t)]\eta(t) \qquad (442)$$

The quadratic noise $\xi^2(t)$ can be written as

$$\xi^2 = \sigma^2 + \Delta\xi \qquad (443)$$

where $\sigma^2 = AB$ and $\Delta = A - B$. For $\xi = A$, one obtains $\xi^2 = AB + (A - B)A = A^2$, and for $\xi = -B$, one obtains $\xi^2 = AB - (A - B)B = B^2$. Therefore, Eq. (442) takes the following form

$$[1 + \sigma^2 + \Delta\xi]\frac{d^2\phi}{dt^2} + \frac{\gamma}{l^2}\frac{d\phi}{dt} + \omega_0^2 \sin\phi = \eta(t) \qquad (444)$$

where, as before, $\langle \xi^2(t_1)\eta(t_2) \rangle = 0$.

This model was originally proposed in the context of a Brownian particle undergoing random adsorption and desorption of surrounded particles [28]. A multiplicative random force arises from the adhesion of surrounding molecules which stick to the Brownian particle for some (random) time, thereby changing its mass. It turns out that the usual condition for Brownian motion — the small size of the surrounding particles compared with the size of the Brownian particle — has to be supplemented by an additional requirement on the mass of Brownian particle for the system to be stable.

The full non-linear problem is described by the following equation

$$[1 + \sigma^2 + \Delta\xi]\frac{d^2\phi}{dt^2} + \frac{\gamma}{l^2}\frac{d\phi}{dt} + \omega_0^2 \sin\phi = a + \eta(t) \qquad (445)$$

where the constant torque a is added to the right side of this equation. A qualitative analysis of Eq. (445) has been performed [131]. For $\xi(t) = \pm\sigma$ with $\langle \xi(t_1)\xi(t_2) \rangle = \sigma^2 \exp(-\lambda|t_1 - t_2|)$ and $\langle \eta(t_1)\eta(t_2) \rangle = \kappa T\gamma\delta(|t_1 - t_2|)$, two Langevin equations

$$[1 + \sigma]\frac{d^2\phi}{dt^2} + \frac{\gamma}{l^2}\frac{d\phi}{dt} + \omega_0^2 \sin\phi = a + \eta(t)$$

$$[1 - \sigma]\frac{d^2\phi}{dt^2} + \frac{\gamma}{l^2}\frac{d\phi}{dt} + \omega_0^2 \sin\phi = a + \eta(t) \qquad (446)$$

correspond to the Fokker-Planck equations

$$\frac{\partial P_\pm}{\partial t} = -\frac{\partial}{\partial\phi}(\Omega P_\pm)$$

$$+ \frac{1}{(1 \pm \sigma)}\frac{\partial}{\partial\Omega}\left(\frac{\gamma}{l^2}\Omega + \omega_0^2 \sin\phi - a + \frac{\kappa T\gamma}{2(1 \pm \sigma)l^2}\frac{\partial}{\partial\Omega}\right)P_\pm$$

$$(447)$$

Equations (447) can be replaced by the set of equations for $P = P_+ + P_-$ and $S = P_+ - P_-$,

$$\frac{\partial P}{\partial t} = -\frac{\partial}{\partial \phi}(\Omega P) + \frac{1}{(1-\sigma^2)}\frac{\partial}{\partial \Omega}$$

$$\times \left[\left(\gamma Q + J\sin\phi - a + \frac{\kappa T\gamma}{2(1-\sigma^2)}\frac{\partial}{\partial \Omega}\right)(P - \sigma S)\right] \equiv \frac{\partial J}{\partial \phi}$$

$$(448)$$

$$\frac{\partial S}{\partial t} = -\frac{\partial}{\partial \phi}(\Omega S) + \frac{1}{(1-\sigma^2)}$$

$$*\frac{\partial}{\partial \Omega}\left[\left(\gamma Q + J\sin\phi - a + \frac{\kappa T\gamma}{2(1-\sigma^2)}\frac{\partial}{\partial \Omega}\right)(S - \sigma P)\right]$$

$$(449)$$

The last two equations have to be solved numerically, and results compared with $\langle d\phi/dt \rangle$ as a function of a for random stiffness and damping [131] and for the correlation function $\langle \frac{d\phi}{dt}(t_1)\frac{d\phi}{dt}(t_2)\rangle$ [132].

3.7 Spring pendulum

The harmonic oscillator is the simplest linear model. However, almost all phenomena in Nature are non-linear. The simplest non-linear generalization of the oscillator model is a pendulum, which consists of bob of mass $m = 1$ suspended by a massless rod of length l, and able to perform plane oscillations about the vertical position (Fig. 10a). During these oscillations, gravity provides a restoring torque $-gl\sin\phi$, which, according to the Newton's second law of circular motion, equals the product of the moment of inertia ml^2 times the angular acceleration $d^2\phi/dt^2$,

$$\frac{d^2\phi}{dt^2} + \omega_0^2\sin\phi = 0 \qquad (450)$$

where $\omega_0^2 = g/l$.

The equation of motion (450) can be obtained from the Lagrangian L written in the polar coordinates R and ϕ,

$$L = \frac{1}{2}\left[\left(\frac{dR}{dt}\right)^2 + \left(R\frac{d\phi}{dt}\right)^2\right] - mg(1 - \cos\phi) \qquad (451)$$

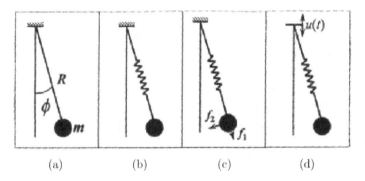

Figure 10: (a) Mathematical pendulum, (b) spring pendulum, (c) spring pendulum subject to an external force, (d) spring pendulum with vertically oscillating suspension point.

Using coordinates R and ϕ, we assume that the motion is planar. For the three-dimensional case, one has three coordinates, R, ϕ, and the angle ψ. The equation of motion for ψ has the following form [133]

$$\frac{d}{dt}\left[R^2 \sin^2 \phi \left(\frac{d\psi}{dt}\right)\right] = 0 \qquad (452)$$

which means that if we initially choose $d\psi/dt = 0$, then $d\psi/dt$ remains zero for all time.

Here we are interested in a pendulum complicated by attributing to the rod a spring of stiffness constant κ shown in Fig. 10b ("spring pendulum"). This implies that one has to add to Lagrangian L the additional potential energy $(\kappa/2)\,(r-l)^2$. Adding the damping term proportional to the first derivatives, one gets the following Lagrangian equations of motion for the extension r of the spring and for the angle ϕ between the spring and the vertical,

$$\frac{d^2 r}{dt^2} + c_1 \frac{dr}{dt} + \frac{\kappa}{m}r - (1+r)\left(\frac{d\phi}{dx}\right)^2 + \frac{g}{l}(1 - \cos\phi) = 0$$

$$(453)$$

$$(1+r)^2\frac{d^2\phi}{dt^2} + c_2\frac{d\phi}{dt} + 2(1+r)\frac{dr}{dt}\frac{d\phi}{dt} + \frac{g}{l}(1+r)\sin\phi = 0$$

$$(454)$$

where $r = R/l - 1$ is the dimensionless length and c_1, c_2 are the damping coefficients. Equations (453) and (454) contain frequencies of two characteristic modes, the oscillator mode $\omega_0 = \sqrt{g/l}$ and the spring mode $\omega_s = \sqrt{\kappa}$, while the nonlinear terms describe the interaction between these two modes.

A further complication of the pendulum model consists of adding an external periodic force of frequency Ω. The latter can be added in two different ways. One way is as an external force acting on the bob which leads to additional terms in Eqs. (453) and (454),

$$\frac{d^2r}{dt^2} + c_1\frac{dr}{dt} + \kappa r - (1 + r)\left(\frac{d\phi}{dt}\right)^2 + \frac{g}{l}(1 - \cos\phi) = K\cos(\Omega t)$$

(455)

$$(1 + r)^2\frac{d^2\phi}{dt^2} + c_2\frac{d\phi}{dt} + 2(1 + r)\frac{dr}{dt}\frac{d\phi}{dt} + \frac{g}{l}(1 + r)\sin\phi = L\cos(\Omega t)$$

(456)

Another possibility, shown in Fig. 10d, corresponding to a periodic force inducing vertical oscillations $u(t)$ of the suspension point, which can be taken into account in the following way. The suspension point has the acceleration d^2u/dt^2 relative to our inertial frame of reference. We introduce a non-inertial frame which has this acceleration. In this accelerated frame, the gravity g in the Lagrangian (451) has to be replaced by $g + d^2u/dt^2$. Therefore, for $u(t) = A\cos(\Omega t)$, Eqs. (453) and (454) have to be replaced by the following equations [134],

$$\frac{d^2r}{dt^2} + c_1\frac{dr}{dt} + \kappa r - (1 + r)\left(\frac{d\phi}{dt}\right)^2 + \frac{g}{l}(1 - \cos\phi)$$

$$- \frac{A\Omega^2}{l}\cos(\Omega t)\cos\phi = 0 \qquad (457)$$

$$(1 + r)^2\frac{d^2\phi}{dt^2} + c_2\frac{d\phi}{dt} + 2(1 + r)\frac{dr}{dt}\frac{d\phi}{dt} + \frac{g}{l}(1 + r)\sin\phi$$

$$+ \frac{A\Omega^2}{l}\cos(\Omega t)\sin\phi = 0 \qquad (458)$$

An external periodic signal $\cos(\Omega t)$ enters Eq. (455) and (456) additively, and Eqs. (457), (458) multiplicatively, in the same way as it was the case in Eqs. (357) and (359) for the simple oscillator. Since the case of an additive external force has been repeatedly discussed (see, for example, [135], [136] and references therein), we will analyze the case of vertical oscillations of the suspension point. We know of only a few papers which considered this problem, such as the stability analysis in a linear approximation using the linear Floquet theory [137], the lowest orders of the method of multiple scales for small amplitude of an external field [138], and the stability analysis of the inverted pendulum [134].

Substituting $\sin\phi \approx \phi - \phi^3/6$ and $\cos\phi \approx 1 - \phi^2/6 + \phi^4/24$ in Eqs. (457)–(458), and keeping only linear terms, one can rewrite these equations in terms of dimensionless time $\tau = \Omega t$

$$\frac{d^2 r}{d\tau^2} + \frac{c_1}{\Omega}\frac{dr}{d\tau} + \frac{\omega_s^2}{\Omega^2}r = \frac{A}{l}\cos\tau \qquad (459)$$

$$\frac{d^2\phi}{d\tau^2} + \frac{c_2}{\Omega}\frac{d\phi}{d\tau} + \left[\frac{\omega_0^2}{\Omega^2} + \frac{A}{l}\cos\tau\right]\phi = 0 \qquad (460)$$

Like (357) and (359), Eq. (459) is the equation of a driven harmonic oscillator while (460) is the Mathieu equation. The detailed analysis of these two equations has been given [134]. These equations have to be compared with the linearized equations describing the externally driven spring pendulum which are simpler than (459), (460), having the form of driven oscillator equations.

Much effort has gone into the study of the pendulum, which is one of the simplest non-linear systems. The method of multiple scales is used for the approximate analytic solution of non-linear problems in different areas of mathematical physics. As a new application of this method, we use it for analysis of the spring pendulum with vertical oscillations of the suspension point and compare the results with those obtained earlier for a driven spring pendulum.

3.8 Analysis of nonlinear equations

3.8.1 *Small external force and damping*

Non-linear equations (457)–(458) cannot be solved analytically. For an approximate solution, we use the method of multiple scales, dividing the motion into the fast regime with $T_0 = t$ and the slow regime with $T_i = \varepsilon^i t, (i = 1, 2 \ldots)$, where $\varepsilon \ll 1$. The time derivatives take the following form

$$\frac{d}{d\tau} = \frac{\partial}{\partial T_0} + \varepsilon \frac{\partial}{\partial T_1} + \varepsilon^2 \frac{\partial}{\partial T_2} + \cdots$$

$$\frac{d^2}{d\tau^2} = \frac{\partial^2}{\partial T_0^2} + 2\varepsilon \frac{\partial^2}{\partial T_0 \partial T_1} + \varepsilon^2 \left(\frac{\partial^2}{\partial T_1^2} + 2\frac{\partial^2}{\partial T_0 \partial T_2} \right)$$

$$+ 2\varepsilon^3 \frac{\partial^2}{\partial T_1 \partial T_2} + \cdots \tag{461}$$

A detailed description of the application of the method of multiple scales for the simplest non-linear equation (Eq. (461) with $\sin \phi$ replaced by $\phi - \phi^3/3$ can be found in a textbook [139]. We start with the case in which the viscous damping force and the external force are equally small, i.e.,

$$c_i \approx \varepsilon; \quad A \approx \varepsilon^2 \tag{462}$$

and expand $\sin \phi \approx \phi - \phi^3/6$, $\cos \phi \approx 1 - \phi^2/2 + \phi^4/24$.

We seek solutions of Eqs. (457)–(458) of the form

$$r = \varepsilon r_1(T_0, T_1) + \varepsilon^2 r_2(T_0, T_1) + \varepsilon^3 r_3(T_0, T_1, T_2)$$

$$+ \varepsilon^4 r_4(T_0, T_1, T_2) + \cdots$$

$$\phi = \varepsilon \phi_1(T_0, T_1) + \varepsilon^2 \phi_2(T_0, T_1) + \varepsilon^3 \phi_3(T_0, T_1, T_2)$$

$$+ \varepsilon^4 \phi_4(T_0, T_1, T_2) + \cdots . \tag{463}$$

Substituting (461)–(463) into (457), (458) and equating the terms of order ε^i $(i = 1 \ldots 4)$, one gets

$$\left(\frac{\partial^2}{\partial T_0^2} + \omega_s^2 \right) r_1 = 0; \quad \left(\frac{\partial^2}{\partial T_0^2} + \omega_0^2 \right) \phi_1 = 0 \tag{464}$$

$$\left(\frac{\partial^2}{\partial T_0^2} + \omega_s^2 \right) r_2 = -2\frac{\partial^2 r_1}{\partial T_0 \partial T_1} - c_1 \frac{\partial r_1}{\partial T_0} - \frac{\omega_0^2}{2} \phi_1^2$$

$$+ \left(\frac{\partial \phi_1}{\partial T_0} \right)^2 + \frac{A\Omega^2}{l} \cos(\Omega T_0) \tag{465}$$

$$\left(\frac{\partial^2}{\partial T_0^2} + \omega_0^2\right)\phi_2 = -2\frac{\partial^2\phi_1}{\partial T_0\partial T_1} - c_2\frac{\partial\phi_1}{\partial T_0} - 2\frac{\partial r_1}{\partial T_0}\frac{\partial\phi_1}{\partial T_0}$$

$$-\omega_0^2 r_1\phi_1 - 2r_1\frac{\partial^2\phi_1}{\partial T_0^2} \qquad (466)$$

$$\left(\frac{\partial^2}{\partial T_0^2} + \omega_s^2\right)r_3 = -2\frac{\partial^2 r_2}{\partial T_0\partial T_1} - \left(\frac{\partial^2}{\partial T_1^2} + 2\frac{\partial^2}{\partial T_0\partial T_2}\right)r_1 - c_1$$

$$\times\left(\frac{\partial r_2}{\partial T_0} + \frac{\partial r_1}{\partial T1}\right) + 2\left(\frac{\partial\phi_2}{\partial T_0} + \frac{\partial\phi_1}{\partial T_1}\right)\frac{\partial\phi_1}{\partial T_0}$$

$$+\left(\frac{\partial\phi_1}{\partial T_0}\right)^2 r_1 - \omega_0^2\phi_1\phi_2 \qquad (467)$$

$$\left(\frac{\partial^2}{\partial T_0^2} + \omega_0^2\right)\phi_3 = -2\frac{\partial^2\phi_2}{\partial T_0\partial T_1} - \left(\frac{\partial^2}{\partial T_1^2} + 2\frac{\partial^2}{\partial T_0\partial T_2}\right)\phi_1$$

$$-\left(c_2 + 2\frac{\partial r_1}{\partial T_0}\right)\left(\frac{\partial\phi_2}{\partial T_0} + \frac{\partial\phi_1}{\partial T_1}\right)$$

$$-2r_1\frac{\partial^2\phi_2}{\partial T_0^2} - 4r_1\frac{\partial^2\phi_1}{\partial T_0\partial T_1} - (r_1^2 + 2r_2)\left(\frac{\partial^2\phi_1}{\partial T_0^2}\right)$$

$$-2\frac{\partial\phi_1}{\partial T_0}\left(\frac{\partial r_2}{\partial T_0} + \frac{\partial r_1}{\partial T_1} + r_1\frac{\partial r_1}{\partial T_0}\right)$$

$$+\frac{\omega_0^2\phi_1^3}{6} - \omega_0^2(r_1\phi_2 + r_2\phi_1) - \frac{A\Omega^2}{l}\cos(\Omega T_0)\phi_1$$

$$(468)$$

$$\left(\frac{\partial^2}{\partial T_0^2} + \omega_s^2\right)r_4 = -2\frac{\partial^2 r_3}{\partial T_0\partial T_1} - 2\frac{\partial^2 r_1}{\partial T_1\partial T_2} - \left(\frac{\partial^2}{\partial T_1^2} + 2\frac{\partial^2}{\partial T_0\partial T_2}\right)r_2$$

$$-c_1\left(\frac{\partial r_3}{\partial T_0} + \frac{\partial r_2}{\partial T_1} + \frac{\partial r_1}{\partial T_2}\right)$$

$$+2\left(\frac{\partial\phi_2}{\partial T_0} + \frac{\partial\phi_1}{\partial T_1}\right)\frac{\partial\phi_1}{\partial T_0}r_1 + \left(\frac{\partial\phi_2}{\partial T_0} + \frac{\partial\phi_1}{\partial T_1}\right)^2$$

$$+ 2\left(\frac{\partial \phi_3}{\partial T_0} + \frac{\partial \phi_2}{\partial T_1} + \frac{\partial \phi_1}{\partial T_2}\right)\frac{\partial \phi_1}{\partial T_0}$$

$$+ \left(\frac{\partial \phi_1}{\partial T_0}\right)^2 r_2 - \omega_0^2\left(\frac{\phi_2^2}{2} + \phi_1\phi_3 - \frac{\phi_1^4}{24}\right)$$

$$- \frac{A\Omega^2}{2l}\cos(\Omega t)\phi_1^2 \tag{469}$$

$$\left(\frac{\partial^2}{\partial T_0^2} + \omega_s^2\right)\phi_4 = -2\frac{\partial^2 \phi_3}{\partial T_0 T_1} - 2\frac{\partial^2 \phi_1}{\partial T \partial T} - \left(\frac{\partial^2}{\partial T_1^2} + 2\frac{\partial^2}{\partial T_0 \partial T_2}\right)\phi_2$$

$$- \left(c_2 + 2\frac{\partial r_1}{\partial T_0}\right)\left(\frac{\partial \phi_3}{\partial T_0} + \frac{\partial \phi_2}{\partial T_1} + \frac{\partial \phi_1}{\partial T_2}\right)$$

$$- 2(r_3 + r_1 r_2)\frac{\partial^2 \phi_1}{\partial T_0^2} - 2r_1\left[\frac{\partial \phi_3}{\partial T_0} + 2\frac{\partial^2 \phi_2}{\partial T_0 \partial T_1}\right.$$

$$+ \frac{\partial^2 \phi_1}{\partial T_1^2} + 2\frac{\partial^2 \phi_1}{\partial T_0 \partial T_2} + \frac{\partial r_1}{\partial T_0}\left(\frac{\partial \phi_1}{\partial T_1} + \frac{\partial \phi_2}{\partial T_0}\right)$$

$$\left. + \frac{\partial \phi_1}{\partial T_0}\left(\frac{\partial r_1}{\partial T_1} + \frac{\partial r_2}{\partial T_0}\right)\right]$$

$$- 2r_2\left(\frac{\partial^2 \phi_2}{\partial T_0^2} + 2\frac{\partial^2 \phi_1}{\partial T_0 \partial T_1} + \frac{\partial r_1}{\partial T_0}\frac{\partial \phi_1}{\partial T_0}\right)$$

$$- r_1^2\left(\frac{\partial^2 \phi_2}{\partial T_0^2} + 2\frac{\partial^2 \phi_1}{\partial T_0 \partial T_1}\right)$$

$$- 2\frac{\partial \phi_1}{\partial T_0}\left(\frac{\partial r_3}{\partial T_0} + \frac{\partial r_2}{\partial T_1} + \frac{\partial r_1}{\partial T_2}\right)$$

$$- 2\left(\frac{\partial r_2}{\partial T_0} + \frac{\partial r_1}{\partial T_1}\right)\left(\frac{\partial \phi_2}{\partial T_0} + \frac{\partial \phi_1}{\partial T_1}\right)$$

$$- \omega_0^2\left(r_1\phi_3 + r_3\phi_1 + r_2\phi_2 - \frac{r_1\phi_1^3}{6} - \frac{\phi_1^2\phi_2}{2}\right)$$

$$- \frac{A\Omega^2}{l}\cos(\Omega t)\phi_2 \tag{470}$$

Equations (464)–(470) give the appropriate solution of equations (457)–(458) which describe the motion of a spring pendulum with vertical oscillations of the suspension point. These equations are similar to those describing motion of a spring pendulum subject to an external force [135, 136, 140]. Let us compare our results with those obtained in these articles. As expected, the difference lies in the position of an external force. Up to order ε^2, both cases are described by the same equations (464)–(466). The difference starts in $O(\varepsilon^3)$ terms in Eq. (470) for the angle ϕ, and in $O(\varepsilon^4)$ terms in Eq. (469) for the length r. The external force appears in these equations for the vertical oscillations, but is absent in the appropriate equations for an external force. However, this difference does not lead to qualitatively new results, since the external force is proportional to $\exp[i(\pm\omega_0 \pm \Omega)T_0]$, and such terms already appear in the appropriate equations in $O(\varepsilon^2)$ and in $O(\varepsilon^3)$.

The solutions of Eqs. (464) have the following form

$$r_1 = C_1(T_1)\exp(i\omega_s T_0) + \widehat{C_1}(T_1)\exp(-i\omega_s T_0)$$
$$\phi_1 = C_2(T_1)\exp(i\omega_0 T_0) + \widehat{C_2}(T_1)\exp(-i\omega_0 T_0) \qquad (471)$$

As an example of succeeding calculations, consider the results of substituting (471) into some of the terms in Eqs. (465)–(470).

$$\phi_1^2 = [C_2\exp(i\omega_0 T_0) + \widehat{C_2}\exp(-i\omega_0 T_0)]^2$$
$$= C_2^2\exp(2i\omega_0 T_0) + C_2\widehat{C_2} + c.c. \qquad (472)$$

$$\left(\frac{\partial\phi_1}{\partial T_0}\right)^2 = -\omega_0^2[C_2^2\exp(2i\omega_0 T_0) - C_2\widehat{C_2} + c.c.] \qquad (473)$$

$$\frac{\partial r_1}{\partial T_0}\frac{\partial\phi_1}{\partial T_0} = -\omega_0\omega_s\{C_1 C_2\exp[(i(\omega_0 + \omega_s)T_0)]$$

$$+C_1\widehat{C_2}\exp[(i(\omega_0 - \omega_s)T_0)] + c.c.\} \qquad (474)$$

where *c.c.* denotes the complex conjugate.

Calculating the other terms in (465), (466) in a similar way converts the latter equations into the following form

$$r_1 = C_1(T_1)\exp(i\omega_s T_0) + \widehat{C_2}(T_1)\exp(-i\omega_s T_0)$$
$$\phi_1 = C_2(T_1)\exp(i\omega_0 T_0) + \widehat{C_2}(T_1)\exp(-i\omega_0 T_0) \qquad (475)$$

$$\left(\frac{\partial^2}{\partial T_0^2} + \omega_s^2\right) r_2 = -i\left(2\frac{\partial}{\partial T_1} + c_1\right) C_1 \omega_s \exp(i\omega_s T_0)$$

$$+ \frac{A\Omega^2}{l} \exp(i\Omega T_0)$$

$$- \frac{\omega_0^2}{2}[3C_2^2 \exp(2i\omega_0 T_0) - C_2\widehat{C_2}] + c.c. \qquad (476)$$

$$\left(\frac{\partial^2}{\partial T_0^2} + \omega_0^2\right) \phi_2 = -i\omega_0\left(2\frac{\partial}{\partial T_1} + c_2\right) C_2 \exp(i\omega_0 T_0)$$

$$+ \omega_0[(2\omega_s + \omega_0)C_1 C_2 * \exp(i(\omega_0 + \omega_s)T_0)$$

$$- (2\omega_s - \omega_0)C_2\widehat{C_1} \exp(i(\omega_s - \omega_0)T_0)] + c.c \qquad (477)$$

Equations (476) and (477) are quite cumbersome. We are interested in the special case of external resonance $\Omega \approx \omega_s$ and internal autoparametric resonance when $\omega_s \approx 2\omega_0$.

To continue the calculation for the resonance case $\Omega \approx \omega_s$ and $\omega_s \approx 2\omega_0$, one has to eliminate the secular terms in the right-hand sides of (476) and (477) by introducing the parameters σ_1 and σ_2 [135, 136],

$$\Omega = \omega_s + \varepsilon\sigma_1; \quad \omega_s = 2\omega_0 + \varepsilon\sigma_2 \qquad (478)$$

Then, Eqs. (476) and (477) can be rewritten in the form

$$\left(\frac{\partial^2}{\partial T_0^2} + \omega_s^2\right) r_2 = \left\{-i\left(2\frac{\partial}{\partial T_1} + c_1\right) C_1 \omega_s + \frac{A\Omega^2}{l} \exp(i\sigma_1 T_0)\right.$$

$$\left. - \frac{\omega_0^2}{2}[3C_2^2 \exp(i\sigma_2 T_0)]\right\} \exp(i\omega_s T_0) + c.c. $$

$$(479)$$

$$\left(\frac{\partial^2}{\partial T_0^2} + \omega_0^2\right) \phi_2 = \left\{-i\omega_0\left(2\frac{\partial}{\partial T_1} + c_2\right) C_2\right.$$

$$\left. - (2\omega_s - \omega_0)\omega_0 C_2\widehat{C_1} \exp(i\sigma_2 T_0)\right\}$$

$$* \exp(i\omega_0 T_0) + \omega_0(2\omega_s + \omega_0)C_1 C_2$$

$$\times \exp(i(\omega_0 + \omega_s)T_0) + c.c. \tag{480}$$

The solutions of Eqs. (476) and (477) have to remain finite, which imposes the condition of removing all secular terms in these equations,

$$2i\omega_s \frac{\partial C_1}{\partial T_1} = -ic_1\omega_s C_1 - \frac{3}{2}\omega_0^2 C_2^2 \exp(i\sigma_2 T_0) + \frac{A\Omega^2}{l} \exp(i\sigma_1 T_0) \tag{481}$$

$$2i\omega_0 \frac{\partial C_2}{\partial T_1} = -ic_2\omega_0 C_2 - \omega_0(2\omega_s - \omega_0)C_2 \widehat{C_1} \exp(i\sigma_2 T_0) \tag{482}$$

Presenting the solutions of Eqs. (481) and (482) in the form

$$C_1 = \frac{a}{2}\exp(i\alpha); \quad C_2 = \frac{b}{2}\exp(i\beta), \tag{483}$$

and using Eq. (471), yields

$$r_1 = a\cos(\Omega t - \gamma_1); \quad \phi_1 = b\cos\left(\frac{\Omega}{2}t + \frac{\gamma_2 - \gamma_1}{2}\right) \tag{484}$$

From (481) and (482), one obtains

$$\frac{\partial a}{\partial T_1} = -\frac{1}{2}c_1 a - \frac{3\omega_0^2}{8\omega_s}b^2\sin\gamma_2 + \frac{A\Omega^2}{2\omega_s l}\sin\gamma_1$$

$$\frac{\partial b}{\partial T_1} = -\frac{1}{2}c_2 b - \frac{1}{4}(\omega_0 - 2\omega_s)ab\sin\gamma_2 \tag{485}$$

$$a\frac{\partial \alpha}{\partial T_1} = \frac{3\omega_0^2}{8\omega_s}b^2\cos\gamma_2 - \frac{A\Omega^2}{2\omega_s l}\cos\gamma_1$$

$$b\frac{\partial \beta}{\partial T_1} = -\frac{1}{4}(\omega_0 - 2\omega_s)ab\cos\gamma_2 \tag{486}$$

where

$$\gamma_1 = \sigma_1 T_1 - \alpha; \quad \gamma_2 = \sigma_2 T_1 - \alpha + 2\beta \tag{487}$$

Equations for γ_1 and γ_2 follow from (486)

$$a\frac{\partial \gamma_1}{\partial T_1} = a\sigma_1 - \frac{3\omega_0^2}{8\omega_s}b^2\cos\gamma_2 + \frac{A\Omega^2}{2\omega_s l}\cos\gamma_1$$

$$ab\frac{\partial\gamma_2}{\partial T_1} = b\left\{a\sigma_1 - \left[\frac{3\omega_0^2}{8\omega_s}b^2 + \frac{1}{2}(\omega_0 - 2\omega_s)a^2\right]\right.$$

$$\left. \cos\gamma_2 + \frac{A\Omega^2}{2\omega_s l}\cos\gamma_1\right\} \tag{488}$$

Equations (484) represent the approximate solutions of the original Eqs. (457) and (458) expressed in terms of the corresponding solutions (481) and (482). The equilibrium, time-independent solutions of Eqs. (485)–(488) define the steady-state solutions of Eqs. (457) and (458). There are different types of such solutions [135]:

1. A harmonic solution, where the only spring mode corresponds to an external force,

$$a_1 = \frac{A\Omega^2}{\sqrt{\omega_s(c_1^2 + 4\sigma_1^2)}}; \quad b_1 = 0;$$

$$\sin\gamma_1 = \frac{c_1}{\sqrt{c_1^2 + 4\sigma_1^2}}; \quad \cos\gamma_1 = -\frac{2\sigma_1}{\sqrt{c_1^2 + 4\sigma_1^2}} \tag{489}$$

For this case, γ_2 remains undetermined.

2. Two second-order subharmonic solutions with

$$a_2 = \frac{2\sqrt{(\sigma_1 - \sigma_2)^2 + c_2^2}}{2\omega_s - \omega_0} \tag{490}$$

which exist if the quadratic equation

$$b_2^2 - Fb_2 + G = 0 \tag{491}$$

has real roots. Here,

$$F = \frac{16\omega_s\left[2\sigma_1(\sigma_1 - \sigma_2) - c_1 c_2\right]}{3\omega_0^2(2\omega_s - \omega_0)};$$

$$G = \frac{16\omega_s^2(c_1^2 + 4\sigma_1^2)(a_2^2 - a_1^2)}{9\omega_0^4} \tag{492}$$

and

$$\sin\gamma_1 = \frac{\omega_s}{A\Omega^2}\left[c_1 a + \frac{3\omega_0^2}{4\omega_s}b^2\sin\gamma_2\right] \quad \sin\gamma_2 = \frac{c_2}{\sqrt{(\sigma_1 - \sigma_2)^2 + c_2^2}} \tag{493}$$

In addition to the periodic solutions (489)–(493) of nonlinear equations (485) and (488), comprehensive numerical calculations have been performed [135], [136]. The dependence of amplitudes a and b on the characteristic of the external field $A\Omega^2$ has been studied [135] at fixed parameters $\omega_s = 1.0$, $\sigma_1 = 0.004$, $\sigma_2 = 0.03$, and $c_1 = c_2 = 0,005$, whereas their dependence on the parameters σ_1 and σ_2 and the damping coefficients c_1 and c_2 was studied [136] at fixed $\omega_s = 1$ and $A\Omega^2 = 0.0055$. With an increase of $A\Omega^2$, the steady-state solutions (489)–(493) disappear, and are replaced by quasi-periodic and chaotic solutions [135]. At some characteristic value of $A\Omega^2$, the system is shown to exhibit Hopf bifurcation and a sequence of period-doubling bifurcations leading to chaos. This result is supported by the calculation of the Poincare map and the Lyapunov exponents. Similar results were obtained [136] as functions of deviations σ_1 and σ_2 from the external and internal resonances, as well as for different values of damping coefficients. In addition, both amplitudes, a and b, show hysteresis as function of σ_1 (or σ_2) at fixed values of other parameters. Furthermore, the amplitudes show jumping behavior at some values of the parameters. Analysis of chaotic regions shows the appearance of multiple attractors, and possible trajectories which depend on the initial conditions.

3.8.2 *Small external force and small damping*

Thus far, we considered the case (462) where both the internal force and the damping are small to the same extent. We now turn to another case, in which the damping is much smaller than external force, namely

$$c_i \approx \varepsilon, \qquad A \approx \varepsilon \qquad (494)$$

Using the method of multiplicative scales, the only distinction in the equations between cases (462) and (494) is the smallness of the term describing the external field. In line with this, the A-term changes from $O(\varepsilon^2)$ to $O(\varepsilon)$ and $O(\varepsilon^3)$ in equations (464), (465), (467) for the radial coordinates r, and appears in Eq. (466) to order ε^2 for the angular coordinate ϕ. We present here the equations of order ε and ε^2,

$$\left(\frac{\partial^2}{\partial T_0^2} + \omega_s^2\right) r_1 = \frac{A\Omega^2}{l} \cos(\Omega T_0); \quad \left(\frac{\partial^2}{\partial T_0^2} + \omega_0^2\right) \phi_1 = 0 \qquad (495)$$

$$\left(\frac{\partial^2}{\partial T_0^2} + \omega_s^2\right) r_2 = -2\frac{\partial^2 r_1}{\partial T_0 \partial T_1} - c_1 \frac{\partial r_1}{\partial T_0} - \frac{\omega_0^2}{2}\phi_1^2 + \left(\frac{\partial \phi_1}{\partial T_0}\right)^2$$

$$(496)$$

$$\left(\frac{\partial^2}{\partial T_0^2} + \omega_0^2\right) \phi_2 = -2\frac{\partial^2 \phi_1}{\partial T_0 \partial T_1} - c_2 \frac{\partial \phi_1}{\partial T_0} - 2\frac{\partial r_1}{\partial T_0}\frac{\partial \phi_1}{\partial T_0}$$

$$- \omega_0^2 r_1 \phi_1 - 2r_1 \frac{\partial^2 \phi_1}{\partial T_0^2} + \frac{A\Omega^2}{l(\omega_s^2 - \Omega^2)} \cos(\Omega T_0)\phi_1$$

$$(497)$$

Equations (495)–(497) have to be compared with the approximate forms of order ε and ε^2 of Eqs. (455) and (456),

$$\left(\frac{\partial^2}{\partial T_0^2} + \omega_s^2\right) r_1 = K \cos(\Omega T_0); \quad \left(\frac{\partial^2}{\partial T_0^2} + \omega_0^2\right) \phi_1 = L \cos(\Omega T_0)$$

$$(498)$$

$$\left(\frac{\partial^2}{\partial T_0^2} + \omega_s^2\right) r_2 = -2\frac{\partial^2 r_1}{\partial T_0 \partial T_1} - c_1 \frac{\partial r_1}{\partial T_0} - \frac{\omega_0^2}{2}\phi_1^2 + \left(\frac{\partial \phi_1}{\partial T_0}\right)^2$$

$$(499)$$

$$\left(\frac{\partial^2}{\partial T_0^2} + \omega_0^2\right) \phi_2 = -2\frac{\partial^2 \phi_1}{\partial T_0 \partial T_1} - c_2 \frac{\partial \phi_1}{\partial T_0}$$

$$-2\frac{\partial r_1}{\partial T_0}\frac{\partial \phi_1}{\partial T_0} - \omega_0^2 r_1 \phi_1 - 2r_1 \frac{\partial^2 \phi_1}{\partial T_0^2} \qquad (500)$$

A comparison of Eqs. (495)–(497) with (498)–(500) shows that the action of an external force on a pendulum and that of vertical oscillation of the suspension point lead to quite different motions of a pendulum already in order ε. Therefore, it is sufficient to analyze Eqs. (495)–(497), which are of order ε and ε^2. The solutions of Eqs. (495)

have the following form,

$$r_1 = C_1(T_1) \exp(i\omega_s T_0) + \frac{A\Omega^2}{l(\omega_s^2 - \Omega^2)} \exp(i\Omega T_0) + c.c.;$$

$$\phi_1 = C_2 \exp(i\omega_0 T_0) + \widehat{C_2} \exp(-i\omega_0 T_0) \qquad (501)$$

Performing calculations similar to (475)–(477), one transforms Eqs. (496) and (497) to the following form

$$\left(\frac{\partial^2}{\partial T_0^2} + \omega_s^2\right) r_2 = -ic_1 \left[C_1\omega_s \exp(i\omega_s T_0) + \frac{A\Omega^3}{l(\omega_s^2 - \Omega^2)} \exp(i\Omega T_0)\right]$$

$$- 2i\frac{\partial C_1}{\partial T_1}\omega_s \exp(i\omega_s T_0)$$

$$- \frac{\omega_0^2}{2} \left[3C_2^2 \exp(2i\omega_0 T_0) - C_2\widehat{C_2}\right] + c.c. \qquad (502)$$

$$\left(\frac{\partial^2}{\partial T_0^2} + \omega_0^2\right) \phi_2 = -i\omega_0 \left(2\frac{\partial}{\partial T_1} + c_2\right) C_2 \exp(i\omega_0 T_0)$$

$$*[\omega_0(2\omega_s + \omega_0)C_1C_2 \exp(i(\omega_0 + \omega_s)T_0)$$

$$- \omega_0(2\omega_s - \omega_0)C_2\widehat{C_1} \exp(i(\omega_s - \omega_0)T_0)]$$

$$+ \left(\omega_0^2 + \frac{2\Omega^3\omega_0}{l(\omega_s^2 - \Omega^2)}\right) \frac{AC_2}{l(\omega_s^2 - \Omega^2)}$$

$$\times \exp(i(\omega_0 + \Omega)T_0)$$

$$+ \left(\omega_0^2 - \frac{2\Omega^3\omega_0}{l(\omega_s^2 - \Omega^2)}\right) \frac{A\widehat{C_2}}{l(\omega_s^2 - \Omega^2)}$$

$$\times \exp(i(\Omega - \omega_0)T_0) + c.c \qquad (503)$$

Equations (502) and (503) are quite cumbersome. We are interested in the special case of the autoparametric resonance with $\omega_s \approx 2\omega_0$. Unlike the previous section, we cannot consider the external resonance $\Omega \approx \omega_s$ in addition to the internal resonance, since the solution of Eqs. (501) becomes infinite for $\Omega \approx \omega_s$. In order to get finite results, one has to increase the damping, i.e., to return to the analysis performed in the previous section. The solutions of

Eqs. (502) and (503) for the case of autoparametric resonance have to remain finite which requires removing all secular terms in these equations,

$$2i\omega_s \frac{\partial C_1}{\partial T_1} = -ic_1\omega_s C_1 - \frac{3\omega_0^2}{2}C_2^2 \tag{504}$$

$$2i\omega_0 \frac{\partial C_2}{\partial T_1} = -ic_2\omega_0 C_2 - \omega_0(2\omega_s - \omega_0)C_2\widehat{C_1} \tag{505}$$

Writing the solutions of Eqs. (504) and (505) in the form

$$C_1 = \frac{a}{2}\exp(i\alpha); \quad C_2 = \frac{b}{2}\exp(i\beta), \tag{506}$$

and using Eqs. (491) and (492), yields

$$r_1 = a\cos(\omega_s + \alpha) + \frac{A\Omega^2}{l(\omega_s^2 - \Omega^2)}\cos(\Omega T_0); \quad \phi_1 = b\cos(\omega_0 + \beta), \tag{507}$$

From Eq. (501), one obtains

$$\frac{\partial a}{\partial T_1} = -\frac{1}{2}c_1 a - \frac{3\omega_0^2}{8\omega_s}b^2\sin(2\beta - \alpha)$$

$$a\frac{\partial \alpha}{\partial T_1} = \frac{3\omega_0^2}{8\omega_s}b^2\cos(2\beta - \alpha)$$

$$\frac{\partial b}{\partial T_1} = -\frac{1}{2}c_2 b - \frac{1}{4}(\omega_0 - 2\omega_s)ab\sin(2\beta - \alpha)$$

$$b\frac{\partial \beta}{\partial T_1} = -\frac{1}{4}(\omega_0 - 2\omega_s)ab\cos(2\beta - \alpha) \tag{508}$$

Performing these transformations leads to the approximate solutions (507) of the original equations subject to the condition (494). The equation for γ_2 follows from (482),

$$ab\frac{\partial \gamma_2}{\partial T_1} = b\left\{a\sigma_2 - \left[\frac{1}{2}(\omega_0 - 2\omega_s)a^2 + \frac{3\omega_0^2}{8\omega_s}b^2\right]\cos\gamma_2\right\} \tag{509}$$

The numerical solution of non-linear equations (508) and (509) in the absence of damping ($c_1 = c_2 = 0$) has been performed [141]. The time dependence of the amplitudes a and b clearly shows the energy transfer between the spring and oscillator modes ("autoparametric resonance").

The spring pendulum is a two-degree-of-freedom non-linear system characterized by two characteristic frequencies, ω_s of the spring mode and ω_0 of the oscillator mode. In the absence of damping, when $\omega_s = 2\omega_0$, this Hamiltonian system performs autoparametric oscillations [142]. This phenomenon was predicted to be Stephenson as early as 1908 [143]. This means that the usually stable downward position of spring pendulum becomes unstable. Being released from equilibrium, the pendulum performs oscillations in which the total energy is transferred back and forth between the spring and pendulum modes. These oscillations disappear in the presence of damping.

An additional characteristic frequency Ω is introduced into the problem by the application of an external periodic force $A \cos(\Omega t)$. This periodic force, which acts on the bob, enters the equations of motion additively. As a result, an additional, external resonance appears when Ω is equal to ω_s or ω_0. We considered a different way of introducing an external periodic force, namely, by applying this force to the suspension point rather than to the bob. Then, the periodic force enters the equation of motion multiplicatively. In both cases, the external periodic force supplies energy to the system while, due to the damping, this energy (or part of it) is transferred to the surrounding medium. We considered two cases. First, the case in which the external force and the viscous damping force are equally small. Up to second order in the small parameter, the solution is the same for both types of external force, allowing both autoparametric and external resonances. However, when the damping is much smaller then the external force, i.e., the energy is accumulated in a system, the behavior is different. In particular, the external resonance is absent for a pendulum with vertical oscillations of the suspension point. In addition to the dynamic systems considered here, a comprehensive analysis of the noisy oscillator and the pendulum has been performed in [90] and [132], respectively.

In addition to the theoretical interest of the behavior of nonlinear two-degrees-of-freedom systems, these systems are widely used as the models for describing different phenomena from the earthquake gas shut-off [144] to the apparatus for defecting oscillations of the weight [145] and for the shaking of chemical reactions [146], among others.

References

[1] Wikipedia, the free encyclopedia.

[2] L. D. Landau and E. M. Lifshitz, *Statistical Physics*, [Pergamon 1980].

[3] M. C. Wang and G. E. Uhlenbeck, Rev. Mod. Phys. **17**, 323 (1945).

[4] K. R. Kim, D. J. Lee, C. J. Kim and K. J. Shin, Bull. Korean Chem. Soc. **15**, 627 (1994).

[5] C. W. Gardiner, *Handbook of Stochastic Methods* [Springer, Berlin, 2004].

[6] S. C. Venkataramani, T. M. Antonsen, Jr., E. Ott, and J. C. Sommerer, Physica D **96**, 66 (1996).

[7] R. Graham, M. Hoihenerbach, and A. Schenze, Phys. Rev. Lett., **48**, 1396 (1982).

[8] W. Kohler and G. C. Papanicolaou, in *Springer Lectures in Physics*, **70** [Springer, New York 1977].

[9] M. Turelli, *Theoretical Population Biology* [Academic, New York, 1977].

[10] H. Takayasu, A-H. Sato, and M. Takayasu, Phys. Rev. Lett., **79**, 966 (1997).

[11] M. Gitterman, Physica A **352**, 309 (2005).

[12] K. Lindenberg, V. Seshadri and B. West, Physica A **105**, 445 (1988).

[13] B. West and V. Seshadri, J. Geophys. Res. **86**, 4293 (1981).

[14] M. Gitterman, Phys. Rev. E **70**, 036116 (2004).

[15] A. Onuki, J. Phys: Condens. Matter **9**, 6119 (1997).

[16] J. M. Chomaz and A. Couairon, Phys. Fluids, **11**, 2977 (1999).

[17] T. E. Faber, *Fluid Dynamics for Physicists* [Cambridge University Press, Cambridge, 1995].

[18] F. Heslot and A. Libchaber, Phys. Scr. **T9**, 126 (1985).

[19] A. Saul and K. Showalter, *Oscillations and Travel Waves in Chemical Systems,* ed. R. J. Field and M. Burger [Wiley, New York, 1985].

[20] M. Gitterman, B. Ya. Shapiro and I. Shapiro, Phys. Rev. B **65**, 174510 (2002).

[21] Li Jing-Hui, Commun. Theor. Phys. (Beijing, China) **50**, 1159 (2008).

[22] H. Calisto, F. Mora and E. Tiraprgui, Phys. Rev E **74**, 022102 (2008).

[23] Li-juan Ning, and Wei Xu, Chinese J. of Physics, **46**, 611 (2008).

[24] L. Gammaitoni, P. Hanggi, P. Jung and F. Marchesonl, Rev. Mod. Phys. **70**, 223 (1998).

[25] A. A. Zaikin, J. Kurths and L. Schimansky-Geier, Phys. Rev. Lett. **85**, 227 (2000).

[26] J. Luczka and J. Sladkowski, Czech. J. Phys. B **39**, 689 (1989).

[27] M. Gitterman, Journal of Phys. Conf. **248**, 012049 (2010); M. Gitterman and V. Klyatskin, Phys. Rev. E **81** 051139 (2010).

[28] M. Gitterman, *The Noisy Oscillator: Random Mass, Frequency, Damping* [World Scientific, Singapore, 2013].

[29] S. Chandrasekar, Rev. Mod. Phys. **15**, 1 (1943).

[30] Yu. M. Ivanchenko and L. A. Zilberman, Sov. Phys. — JETP **55**, 2395 (1968).

[31] K. Mallick and P. Markcq, Phys. Rev. E **66**, 041113 (2002).

[32] K. Mallick and P. Markcq, J. Phys. A **37**, 4769 (2004).

[33] Yu. B. Simons and B. Meerson, arXiv : 0907.2518 cond-mat, stat-tex.

[34] I. I. Fedchenia and N. A. Usova, Z. Phys. B **50**, 263 (1983).

[35] V. Berdichevsky and M. Gitterman, Phys Rev E **56** 6340 (1997).

[36] C. J. Wang, S. B. Chen and D. C. Mei, Phys Lett A 352 , 119 (2006).

[37] L. Cao and D. J. Wu. Phys. Rev. E 62, 7478 (2000).

[38] M. Gitterman, J. Phys. A **32**, L293 (1999).

[39] R. Mazo, *Brownian Motion: Fluctuations, Dynamics, and Applications* [Oxford Science Publication, 2002].

[40] A. Ishimaru, *Wave Propagation and Scattering in Random Media* [IEEE Press, 1999].

[41] R. Kubo, *Stochastic Processes in Chemical Physics* ed. K. E. Shuler [Wiley, New York, 1969].

[42] J. M. Phillips, *The Dynamics of the Upper Ocean* [Cambridge University Press, 1977].

[43] I. Goldhirsch and G. Zanetti, Phys. Rev. Lett., **70**, 1619 (1993).

[44] M. Gitterman, and D. Kessler, Phys. Rev. E. **87**, 022137 (2013).

[45] T. Kaminski, J. P. Siebrasse, V. Gieselmann and J. Kappler, Glycoconjugate Journal, **25**, 555 (2008).

[46] M. Sewbawe Abdalla, Phys. Rev. A, **34**, 4598 (1986).

[47] J. Portman, M. Khasin, S W Shaw and M I. Dykman, Bulletin of the APS, March Meeting, (2010).

[48] J. Luczka, P. Hanggi and A. Gadomski, Phys. Rev. E, **51**, 5762 (1995).

[49] R. Lambiotte and M. Ausloos, Phys. Rev. E, **73**, 011105 (2005).

[50] A. Gadomski and J. Siódmiak, Cryst. Res. Technol. **37**, 281 (2002).

[51] A. Gadomski, J. Siódmiak, I. Santamarìa-Holek, J. M. Rubì, and M. Ausloos, Acta Phys. Pol. B, **36**, 1537 (2005).

[52] A. T. Pérez, D. Saville, and C. Soria, Europhys. Lett. **55**, 425 (2001).

[53] See I. Temizer, M.Sc. thesis, University of California, Berkeley, www.me.berkeley.edu/compmat/ilkerDOCS/MSthesis.pdf. (2003).

[54] W. Benz, Spatium, **6**, 3 (2000).

[55] S. J. Weidenschilling, D. Spaute, D. R. Davis, F. Marzari, and K. Ohtsuki, Icarus, **128**, 429 (1997).

[56] N. Kaiser, Appl. Opt. **41**, 3053 (2002).

[57] T. Nagatani, J. Phys. Soc. Jpn. **65**, 3386 (1996).

[58] E. Ben-Naim, P. L. Krapivsky, and S. Redner, Phys. Rev. E, **50**, 822 (1994).

[59] Ausloos M. and K. Ivanova, Eur. Phys. J. B **27**, 177 (2002).

[60] Ausloos M. and K. Ivanova, in *Proceedings of the Second Nikkei Econo-physics Symposium*, ed. H. Takayasu, Springer Verlag, Berlin, 2004.

[61] R. Dean Astumian and F. Moss, Chaos **8**, 533 (1998).

[62] W. Horsthemke and R. Lefever, *Noise-Induced Transitions* [Springer, Berlin, 1084].

[63] P. Hanggi and R. Bartussek, *Lecture Notes in Physics* Vol. 476 [Springer, Berlin, 1996].

[64] J M R. Parrondo, C. Van den Broeck and F. de la Rubia, Physica A, **224**, 153 (1996).

[65] P. Yung and P. Talkner, Phys. Rev. E, **51**, 2640 (1995).

[66] I. Dayan, M. Gitterman and G. H. Weiss, Phys. Rev. A, **46**, 757 (1992).

[67] S. L. Ginzburg and M. A. Pustovoit, Phys. Rev. Lett. **80**, 4840 (1998).

[68] C. R. Doering and J. C. Gadoua, Phys. Rev. Lett. **69**, 2318 (1992).

[69] F. Marchesoni, Phys. Lett. A, **237**, 126 (1998).

[70] D. W. Brown, L. J. Bernstein and K Lindenberg, Phys. Rev. E, **54**, 6966 (1996).

[71] G. Nicolis and I. Prigogine, *Self-Organization in Non-Equilibrium Systems* [Wiley, New York, 1977].

[72] J. E. Fletcher, S. Havlin and G. H. Weiss, J. Stat. Phys., **51**, 215 (1988).

[73] P. Reimann and P. Hanggi, *Lecture Notes in Physics* Vol. 484 [Springer, Berlin, 1997].

[74] C. Zhou and C-H Lai, Phys. Rev. E, **59**, R6243 (1999).

[75] P. Yung and E. Marchesoni, Chaos **21**, 046516 (2011).

[76] M. Gitterman, Eur. J. Phys. **23**, 119 (2002).

[77] R. N. Mantegna and R. Spagnolo, Phys. Rev. Lett. **76**, 563 (1996).

[78] L. Landau and E. Lifshitz, *Mechanics*, [Pergamon, 1976].

[79] M. R. Young and S. Singh, Phys. Rev. A **38**, 238 (1988).

[80] Ya Jia and Jia-rong Li, Phys. Rev. E **53**, 5786 (1996).

[81] J. Cveticanin, J. Serbian Soc. for Comp. Mech. **6**, 56 (2012).

[82] V. E. Shapiro and V. M. Loginov, Physica A **91**, 563 (1978).

[83] http://us-mag5.mail.yahoo.com/neo/launch?.rand=5hrgugv.1v903; http://www.sosmath.com/algebra/factor/fac12/fac12.html.

[84] M. Gitterman, Physica A **395**, 119 (2014).

[85] L. Arnold, *Random Dynamic Systems* [Springer, Berlin, 1998].

[86] N. Leprovost, S. Aumaitre and K. Mallick, Eur. Phys. J. B, **49**, 453 (2006).

[87] K. Kitahara, W. Horsthemke, R. Lefever and Y. Inara, Progr. Theor. Phys. **64**, 1233 (1980).

[88] R. C. Bourret , C. H. Frish and A. Pouquet, Physica A **65**, 303 (1973).

[89] S. Roul, in *Programming for Mathematicians* [Springer-Verlag, Berlin, 2000], §10.13.

[90] M. Gitterman, *The Noisy Oscillator: The First Hundred Years from Einstein Until Now* [World Scientific, 2005].

[91] E. Soika, R. Mankin and N. Lumi, AIP conference **1487**, 233 (2012).

[92] M. Gitterman and D. Kessler, Proc. AIP conf. **1558**, 1940 (2013).

[93]   V. I. Klyatskin, Physics-Uspekhi **51**, 395 (2008); **52**, 514 (2009).

[94]   V. I. Klyatskin, *Stochastic Equations through the Eye of the Physicist (Basic Concepts, Exact Results, and Asymptotic Approximations)* [Elsevier, Amsterdam, 2005].

[95]   A. Gadomski and J. Luczka, Fractals, **1**, 875 (1993).

[96]   J. Luczka, P. Hanggi and A. Gadomski, Phys. Rev. E **51**, 5762 (1995).

[97]   M. Gitterman, Physica A **391**, 5343 (2012).

[98]   R. Benzi, S. Sutera, and A. Vulpani, J. Phys. A **14**, L453 (1981).

[99]   G. Nicolis, Tellus, **34**, 1 (1982).

[100] A. Fulinski, Phys. Rev. E **52**, 4523 (1995).

[101] S.-Q. Jiang. B. Wu. and T.-X. Gu, J. Electr. Sci. China, **5**, (4) (2007).

[102] S. Jiang, F. Guo, F, Zhow and T-X. Gu in Communications, Circuits and Systems, ICCCAS (2007), page 1044.

[103] R. L. Lang, L. Yang, H. L. Qin and G. H. Di, Nonlinear Dynamics, **69**, 1423 (2012).

[104] V. I. Klyatskin, Radiophys. Quant. Electr. **20**, 381 (1977).

[105] V. Berdichevsky and M. Gitterman, Phys. Rev. E **60**, 1494 (1999).

[106] F. Sasagawa, Progr. Theor. Phys. **69**, 790 (1983).

[107] K. Ouchi, T. Horita and H. Fujisakqa, Phys. Rev. E **74**, 031106 (2006).

[108] Y. Jia, X-P Zheng, X-M Hu and J. R. Li, Phys. Rev. E **63**, 031107 (2001).

[109] S. Z. Ke, D. J. Wu and L. Gao, Eur. Phys. J. B **12**, 119 (1999).

[110] M. Gitterman, *The Noisy Oscillator* [World Scientific, 2005]; *The Noisy Pendulum* [World Scientific, 2008]: *The Chaotic Pendulum* [World Scientific, 2010].

[111] M. I. Dykman, D. G. Luchinsky, R. Mandella, P. V. E. Mc Clintock, N. D. Stein, N. G. Stocks, Il Nuovo Chimento D **17**, 661 (1995).

[112] A. Fulinski, Phys. Rev. E **52**, 4523 (1995).

[113] M. Gitterman, J. of Phys. C: Conference Series, **248** (1), 012049 (2010): M. Gitterman, J. Stat. Phys. **146** (1), 239 (2010) ; M. Gitterman, J. Modern. Phys. **2**, 1136 (2010).

[114] Y. Braiman and I. Goldhirsch, Phys. Rev. Lett. **66**, 2545 (1991).

[115] Y. Kim, S. Y. Lee and S. Y. Kim, Phys. Lett. **69**, 2318 (1992).

[116] S. Rajasekar, S. Jeyakuman, Y. Chinnanthambi, M. A. Sanuan, J. Phys. A **43**, 465101 (2010).

[117] I. I. Blekhman and P. S. Landa, Int. J. Nonlin. Mech. **39**, 421 (2004).

[118] J. P. Baltanas, L. Lopez, I. I. Blekhman, P. S. Landa, Int. J, Nonlin. Mech. **39**, 421 (2004).

[119] V. N. Chizhevsky, E. Smeu, G. Giacomelli, Phys. Rev. Lett. **91**, 220602 (2003).

[120] S. Jeyakumari, V. Chinnathambi, S. Rajasekar and M. A. F. Sanjuan, Phys. Rev E **80**, 046608 (2009) ; Chaos **19**, 043128 (2009).

[121] J. C. Chedjou, H. B. Fotsin and P. Woafo, Physics Scripta, **55**, 390 (1997).

[122] E. Ott, *Chaos in Dynamic Systems* [Cambridge University Press, 2002].

[123] M. Moshinski and Yu. F. Smirnov *The Harmonic Oscillator in Modern Physics* [Harwood, Netherlands, 1996].

[124] G. L. Baker, Am. J. Phys. **74**, 482 (2006).

[125] K. K. Likharev, *Dynamics of Josephson Junctions and Circuits*, [Gordon and Breach, 1986].

[126] L. R. Nie, D. C. Mei, X. M. Lv, X.X. Sun and P. Li, J. Stat, Phys. **143**, 532 (2011).

[127] V. Berdichevsky and M. Gitterman, Phys. Pev. E **56**, 6349 (1997).

[128] S. Park, S. Kim and C. Ryu, Phys. Lett. A, **225**, 245 (1997).

[129] H. G. Schuster, *Deterministic Chaos* [Weinhelm: Physik-Verlag, 1984].

[130] J. B. Sokolov, Phys. Rep. B **22**, 5832 (1980).

[131] M. Gitterman, *The chaotic pendulum* [World Scientific, 2010], pp. 88–89.

[132] M. Gitterman, *The noisy pendulum* [World Scientific 2008], pp. 47–48.

[133] P. Lynch, Int. J. Nonlin. Mech. **37**, 345, (2002).

[134] A. Arinstein and M. Gitterman, Eur. J. Phys. **29**, 1, (2008).

[135] W. K. Lee and H. D. Park, Nonlin. Dyn. **14**, 211, (1997).

[136] A. Alasty and R. Shabani, Nonlin. Analysis **7**, 81, (2006).

[137] G. Ryland and L. Meirovich, J. Sound. Vibr. **51**, 547, (1977).

[138] C. E. N. Mazzilli, IMA J. Appl. Math. **34**, 137, (1985).

[139] A. H. Nayfeh and D. T. Mook, *Nonlinear Oscillations* [Wiley, New York, 1979].

[140] M. Eissa, S. A. El-Sefari, M. El-Sheikh, and M. Sayed, Appl. Math. Comput. **145**, 421, (2003).

[141] R. Starosta and J. Awrejcewicz, *Proc. 9th Conference on Dynamic Systems*, Lodz, Poland, 2007.

[142] A. A. Vitt and G. S. Gorelik, Zh. Tekh. Fiz. Sovietunion **3**, 291, (1933) (in Russian).

[143] A. Stephenson, Mem. Proc. Manch. Lit. Phil. Soc. **52**, 1, (1908).

[144] US patent 578791.

[145] S. Toyotomi, Phys. Educ. **29**, 382, (1994).

[146] V. A. Lamb and M. M. Harris, J. Chenm. Phys. **17**, 577, (1940).

Index

Printed in the United States
By Bookmasters